About Brady G. Wilson

Brady G. Wilson is, undisputedly, the embodiment of focused energy. As co-founder of Juice Inc., Brady's vision is to create a world where businesses pulsate with creative energy. For 20+ years, he has inspired and energized leaders, managers, and frontline workers in many of North America's Fortune 500 companies. His passion for creating breakthroughs for companies has spawned such innovative tools and programs as The Power of Conversation™, Beyond Engagement™, and The Energy Check™. Brady lives in Guelph, Ontario. He is also the author of three other books dedicated to improving employee performance and business results.

BEYOND ENGAGEMENT

Also by Brady G. Wilson

*JUICE: The Power of Conversation –
The Secret to Releasing Your People's
Brilliance and Expanding Your Leadership*

*FINDING THE STICKING POINT:
Increase Sales by Transforming Customer
Resistance into Customer Engagement*

*LOVE AT WORK:
Why Passion Drives Performance
in the Feelings Economy*

BEYOND ENGAGEMENT

A Brain-Based Approach That Blends the Engagement Managers Want with the Energy Employees Need

BRADY G. WILSON

Foreword by David Zinger

BPS
books

Toronto & New York
www.bpsbooks.com

Published in 2015 by
BPS Books
Toronto and New York
www.bpsbooks.com
A division of Bastian Publishing Services Ltd.

ISBN 978-1-77236-017-2 (paperback)
ISBN 978-1-77236-018-9 (ePDF)
ISBN 978-1-77236-019-6 (ePUB)

Cataloguing-in-Publication Data available from Library
and Archives Canada.

Cover: Gnibel
Text design: Daniel Crack, Kinetics Design, www.kdbooks.ca

*I dedicate this book to all my grandchildren.
You energize my life in more ways than you will ever imagine.
Grandpa's working hard to create a world that will energize you.*

Contents

Foreword

Countless organizations around the globe have failed to see improvement in employee engagement over the past twenty-five years. Other organizations have increased engagement yet employees are drained and depleted.

Are you ready to embark on a refreshing run through the weekly, monthly, and yearly marathon of work? I encourage you to partner with Brady Wilson, an energy architect and Boston Marathon runner, as he gives you perspective, approaches, neuroscience evidence, stories, and practical wisdom to increase engagement and energy where you work.

As Brady shows in this book, we must not be deterred by tension, as tension can be a vital source of energy. We need to move beyond engagement surveys and scores to uncover the bigger backstory that influences performance, results, and relationships at work. Brady teaches us to move in to tension and backstory with skillful conversations.

I like the book's quick and cogent approach. It is not a huge volume to ponder, and then shelve, without taking action. It is an inspirational and practical book infused with stories and energy. Each chapter begins with brain science, how the idea shows up at work and why the chapter matters.

In one chapter, Brady shares his experience of running the Boston Marathon and receiving phenomenal energizing support from the people who came out to watch. Use this book to run along with Brady, not on a marathon but past the organizational, interpersonal, and brain convolutions that can deter, deplete, and disengage. With Brady as your partner, you will reach your finish-line with increased energy, stronger relationships, and robust results.

–DAVID ZINGER
Founder, The Employee Engagement Network

Introduction

After twenty years of trying to get it right, precious few organizations have cracked the code of employee engagement.

Why is this so? Because few of us could have anticipated the unbending nature of what I call "the engagement paradox": the more companies focus on engagement, the more *disengagement* they produce.

What causes this paradox? Simply this: managing engagement turns out to be just another drain on the most precious resource in business today – energy. It's not that employees don't want to be engaged. It's not that they don't support the engagement program. They are committed and loyal soldiers. However, in today's exhaustion era, they are simply struggling to make it to the weekend. Lacking energy, they resort to quick fixes, workarounds, and reactive firefighting, thereby hardwiring depletion into the system.

As a result, employees come to perceive engagement efforts as a management con game. We've witnessed situations where up to 50 percent of the employee population believe no meaningful outcomes will occur as a result of the engagement survey. This crisis of belief causes acute pain inside well-intentioned leaders who are doing their best to unlock employee engagement. They feel caught.

One manager put it aptly when he said, "We talk about engagement all the time. It's like being in a relationship and constantly being asked to talk about the relationship – rather than taking the time to hold hands and walk down the beach."

This engagement paradox is a serious concern, for two reasons.

- First, because, while this engagement-cum-disengagement manifests itself throughout companies' general employee population, it has its greatest impact on the two groups that matter most: high-performers and managers.
- And second, disengagement *misaligns* the vital connection between a great employee experience, a great customer experience, and great results.

How Did We Get Here?

Leaders are often unclear about the reason they chose to pursue employee engagement. It may be helpful for you to step back and take stock of why your organization went down the engagement path. The narrative probably went something like this:

CEO: "I want us to be on the Top 50 Best Employers list two years from now."

COO: "Makes sense, and we'll be able to tap in to more of our employees' discretionary effort to drive efficiencies."

EVP Sales: "I think engagement will give our people an injection of motivation to go the extra mile for our customers."

EVP HR: "I get all of that and I think it's important that we have a way of creating more wellness to reduce the number of stress leaves we're seeing."

EVP Communications: "There isn't a serious organization out there that isn't doing employee engagement. If we don't do this, we're not even in the talent game!"

CIO: "Huh?" (He only heard half of what was being said; he was busy on his laptop.)

Because of our self-fueling engagement approach, Juice has often been consulted by senior teams who are struggling with a stalled engagement process. Perhaps you can relate to the story of many leaders who have partnered with us over the past decade, one of whom said to us:

> The first year was easy – we just addressed the "low-hanging fruit" issues and our employees told us: "We love this. You listened to us – and you did something about it." But each year, responding with visible, concrete action has become more challenging. And in the last few years, our engagement scores have plateaued. Cynicism is starting to breed. Our response rates are going down every year, and over half of our respondents don't believe any meaningful action will happen as a result of this survey.

You might be surprised, but one of the most common organizational responses to an employee engagement survey is...*none at all*. After employees fill out the survey, it's as if all the data go into a black hole. Months pass before anyone finds out the results – and in some organizations, they never do.

I have read thousands of written survey comments, offered by employees who clearly cared enough to invest the time to write. For example: "I don't know why I take the time to fill this thing out. Nothing happens as a result. I love what I do but I'm so frustrated about the barriers that are put in my way. Will anybody listen and take action this time?"

In our conversations with senior leaders, they often confess: "We don't want to turf the whole engagement thing – that would send all the wrong signals. But it clearly is not working. It's hurting morale, fostering cynicism, and corroding goodwill."

At this point, they look us in the eyes and ask: "So what are your recommendations?"

The Brain and Energy

There is a way to get beyond this kind of self-defeating engagement, a way to make engagement work, we tell them, and it is by
managing energy rather than engagement.

Energy has been my predominant theme since I co-founded
Juice Inc. with Alex Somos in 2003, a theme that has been confirmed
by recent discoveries in brain science. Chief among them is that
the executive function of the brain – the power-driver of value-
creation and innovation – runs on energy, not on the dedication
and commitment that engagement policies seek to create.

The importance of employee energy is finally becoming main-
stream. That's why the big survey houses are starting to insert an
energy factor into their employee engagement surveys.

I am not a neuroscientist. I am an energy architect. For more
than two decades I've grappled with and found ways to address the
issues that short-circuit employee engagement. I work with senior
leadership teams to build the conditions in which energy can
flourish and be sustainable in their organizations. I am obsessed
with researching the theories and findings of brain science. This
has deepened my practice and grounded my writing. My purpose
in this book, however, is not to enlighten you with scientific facts,
though I do point you to those facts (more on this, below). Rather,
it is to equip you with an understanding of the amazing results of
managing energy.

What's in This Book

I devote a chapter each to ten leadership principles based on the
findings of brain science. These principles are the key to moving
beyond engagement. They are:

1 Manage Energy, Not Engagement

Brain science shows that when we're low on energy, the first thing
we lose is our executive function and, with it, the power tools of
prioritization, planning, decision-making, self-regulation, and

intuition. By managing energy, you safeguard people's executive function. You unlock the energy that fuels the passion, innovation, and enthusiasm that generate true and sustainable engagement.

2 Deliver Experiences, Not Promises

Our brains pursue activities that promise reward regardless of whether the reward is delivered. Engagement activities that promise much but deliver little corrode the employee experience. As you deliver experiences (not just promises) to employees, you have the chance to create a powerful legacy: employees who say, "Those were the best years of my life."

3 Target Emotion, Not Logic

Managing energy in human-to-human relations can be mapped inside the brain. Energy is generated electrochemically as high-performance hormones are released through quality conversations. And those conversations are most powerful when they deal with emotional experiences, not rational ones. In fact, research by the Corporate Leadership Council shows that tapping in to emotional engagement allows employees to offer you 400 percent more of their discretionary effort.

4 Trust Conversations, Not Surveys

Energy in the workplace is generated primarily through quality conversations. They release ease hormones, trust hormones, focus hormones, and creativity hormones in our brains. These hormones deepen relationships and unlock easier results. Through frequent face-to-face Energy Checks you can draw out the best intel about the lived experience of your employees and unlock the energy that enables them to offer their best stuff.

5 Seek Tension, Not Harmony

Although the brain requires tension to do its best thinking, it perceives it as a threat to be avoided. Leaders typically try to avoid tension by overpowering it, giving in to it, or smoothing it over. But all of these responses forfeit the energy and innovations that exist within tension. Emotional experiences are not difficult to locate. They reside within tension – the gap between what employees believe is possible and their experience of their current reality. Your job is to unlock the power of these poignant experiences by stepping in to the tension and harmonizing competing needs.

6 Practice Partnering, Not Parenting

Our emotional brain perceives shared responsibility as a threat and triggers us to become under- or over-responsible. Leaders usually try to resolve the tension with a parenting approach, overpowering employee concerns or accommodating them. As you shift from parenting to partnering, however, you co-author powerful solutions that employees are willing to adopt and implement. This lifts your burden and gives you back the time and mind-space you need to drive the business.

7 Pull Out the Backstory, Not the Action Plan

Brains thrive on connections, and, in organizations, connections are realized through conversations. Leaders do not have to look long and hard to find partnering opportunities: they are staring up at them from the pages of the last employee engagement report. As, through conversations, you draw out the backstory behind the engagement scores, you can identify what matters most to employees and partner with them to create solutions that will generate meaningful energy. A far better approach than aggressive action plans and solutions that fail to engage employees in conversation.

8 Think Sticks, Not Carrots

Our brains respond more to sticks than to carrots. Leaders often gravitate to offering carrots – recognition, cheerleading, and inspiration – instead of thinking sticks, that is, looking for and addressing the psychological forms of interference that undo employees' best efforts. You will see stunning gains when you remove interference.

9 *Meet Needs, Not Scores*

Our brains make decisions for emotional reasons and then justify them with rational ones. Employees have emotional needs that drive their decisions. When these needs go unmet, they act out in unskillful ways that permeate the organization with interference. As you inspire and sustain energy in your organization by meeting employees' needs, you will buoy their efforts – and save yourself precious time.

10 *Challenge Beliefs, Not Emotions*

Our brains do not allot us the resources to do something until convinced it is possible. Employee beliefs can produce low levels of self-efficacy (powerlessness) that bring engagement initiatives to a standstill. By recalibrating unhelpful beliefs, you will produce a greater sense of agency in your employees.

How to Use This Book

Beyond Engagement is constructed with busy leaders in mind. You can grasp the essence of the ten ideas simply by reading the *what – how – why* section at the beginning of each chapter:

- What's the brain science?
- How does this show up at work?
- Why does it matter?

There is a logical flow to the ideas in this book from beginning to end, but you can start with any chapter that grabs your interest. I have concluded each chapter with two elements: a **Case Story** illustrating the amazing results of managing energy and an **Energy Management Question** intended to spark dialogue between you and your fellow leaders.

Outcomes from Reading This Book

1 *Managing the Whole Employee*

A central theme of *Beyond Engagement* is that we have both an emotional and a rational brain and that when leaders learn how to partner with both brains, they release their employees to do remarkable things. By reading this book, you will be enabled to work with the whole person and see the remarkable results that flow from humanizing the workplace.

2 Aligning the Employee Experience and Customer Experience

Early in this introduction I referred to the customer experience, the improvement of which is the ultimate goal of organizational engagement efforts. By managing energy, you will be able to truly align employee experience and customer experience. Energized employees are engaged employees, and engaged employees create engaged customers.

3 Putting Engagement in the Hands of Employees

We've seen leaders achieve amazing things by moving beyond engagement work: some have taken their departments and their organizations from the bottom of the heap to the top. They've done that by placing engagement in the hands of the one person who is never asked to own it – the employee – which is where the responsibility should have been all along.

Did you really sign up to manage people's engagement, acting like a cheerleader, a spinmeister, a social convener? Your job is not to inspire people to get on the engagement bandwagon. It's not to get people to try harder. It's to partner with them to help them manage their own engagement.

4 Achieving Organizational Sustainability

What's the bottom line? It's this: *Engagement is an outcome, not a strategy.* As we at Juice have seen over and over again, leaders who manage energy find their *own* energy replenished. And energized leaders are uniquely positioned and empowered to build a strong and ongoing bridge between their organization and their customers.

If that sounds good to you, read on.

1

Manage Energy, Not Engagement

What's the brain science? When our brain is low on energy, the first thing we lose is our executive function (EF), by which we process, predict, prioritize, and plan.

How does this show up at work? Typical engagement initiatives don't focus on generating energy but on unlocking discretionary effort – getting people to go above and beyond the call of duty. So good-hearted employees heed the call, come in earlier, stay later, and try harder in between. *But effort without innovation just won't cut it.*

Loyal, committed people make Herculean dives to make sure things don't slip through the cracks, but they don't think innovatively about how to get to root causes and fix systemic issues. The result? A culture that rewards firefighting and reactivity: the perfect ecosystem for depletion.

Why does this matter? Workarounds and firefighting perpetuate a cycle of depletion that looks like this:

1 people have even less energy to innovate,
2 which produces even more reactive behaviors,

3 which allows more entropy and decay – guaranteeing further depletion.

Shifting from managing engagement to managing energy changes this game. It fuels both innovative thinking and discretionary effort. This blend of intelligent effort is the only way to sustain energy and drive results.

▶ ▶ ▶

The Gas Guzzler

Consider your brain, that three-pound tub of tofu nestled inside your cranium. It comprises a mere 2 percent of your body weight, but how much energy does it consume? A stunning 20 percent!

Michael Galliot, among others, has shown how the brain is one of the most fuel-hungry organs in the human body. That makes sense, because it houses the executive function (EF), a central processing unit capable of astonishing levels of value creation.

The EF offers four special capabilities that enable you to:

1 Process: analyze, predict outcomes, and problem-solve
2 Focus: memorize, pay attention, and verbalize
3 Self-regulate: maintain impulse control, self-monitor, and cognitively flex
4 Initiate: prioritize, plan, and decide

Predict how productive you'd be on a day if your EF were shut down. What would be left if you didn't have the capacity to process, focus, self-regulate, or initiate action? Reptilian responses, that's what, each one devoid of future-based thought.

When the body's energy tank runs low, the brain prioritizes the use of fuel, giving first dibs to "survival equipment":

- Autonomic responses (blinking, breathing, and pulse)
- Immune and digestive systems
- Unconscious thoughts
- Balance and locomotion
- Fight/flight safety features

The body's use of fuel is judicious: it considers safety a necessity and self-actualization a luxury. When you are low on energy, your base-level thinking continues to function, but your mind's power tools fail to operate.

For example, a study done by the University of London and reported by CNN discovered that the depletion caused from always being "on" – running from meeting to meeting, hyper-vigilantly text messaging and emailing – reduces your IQ by an average of 10 points. That's the equivalent of missing a night's sleep (and, for men, about three times the effect of smoking marijuana).

Knowledge workers need well-fueled brains, because the executive function is what enables them to:

- Think strategically: addressing systemic issues, uncovering root causes, and predicting the downstream implications of decisions and actions
- Collaborate broadly: influencing and aligning stakeholders across the organization
- Communicate clearly: providing context, making meaning, harmonizing competing priorities, and resolving conflict
- Execute decisively: drawing out the best information possible, making a call, and closing all the loops to ensure complete follow-through has been achieved

Just think about your own energy level. How energized are you to perform these activities?

Say you're on your way to work and you're thinking, "I need to have that tough conversation with Ellen" (resolving conflict). If you already feel run down and low on energy, you will probably say to yourself, "I'll do it tomorrow."

When you get to the office and sit down at your desk and pull up that high-level document you need to write (which requires thinking creatively and strategically), you feel depleted.

What will you do instead? Probably check your email, or watch a YouTube video, or organize your office, or drop in on a co-worker

– *anything* but one of those value-adding activities that requires so much of your energy.

Given the choice of whether to "generate energy" or "get stuff done," most organizations and managers ignore the former and obsess about the latter. As a result, our workplaces are filled with people who are engaged but not energized. Here's a story that illustrates the impact of being engaged but not energized.

A Tale of Two Marathons

Participating in two Boston Marathons taught me the crucial difference between engagement and energy.

I ran the first one in 2012. The race organizers had warned it would be an unseasonably warm day and gave us the opportunity to defer to 2013.

But I had trained hard, driven for almost ten hours, and paid a handsome price for my hotel room. I felt a strong sense of compulsion, obligation, and gritty perseverance. I was going to run this race.

I barely finished. It took a grueling four hours and thirty minutes. There were times I thought I'd have to quit, but somehow I soldiered on to the end. The best summary I can give is that I was engaged but not energized.

Despite my struggle, I fell head-over-heels in love with Boston that year: the city, the devoted spectators, my fellow runners from around the world, and the Boston Athletic Association, which put on such a phenomenal race. Still, I said to myself, "I'm never doing another marathon again – I'm done with that."

And then, in 2013, the bombs went off.

My friend Stan was running in the marathon that fateful day, and his typical completion time would have put him right at the finish-line when the first bomb exploded. I was frantic, madly texting him to see if he was safe. I breathed again when he texted us back, "I'm OK. I crossed the line and heard an explosion."

In the ensuing hours, I witnessed the horror: twenty-two

victims, runners and spectators alike, mutilated by this senseless act. This may sound odd to you, but what evoked such a strong empathetic response in me was the nature of their injuries. Most of the victims had severe leg injuries; as someone who loves to run – lives to run – this is what gripped my heart.

Something rose up within me in that moment – the same thing that rose up in thousands of runners around the world: "I am so running Boston next year. I will not be intimidated by this." I felt huge solidarity with the victims and their families. I felt huge solidarity with the city of Boston. I felt huge solidarity with the Boston Athletic Association, whose pristine race had been ravaged by this senseless act.

So in the following month, with little time to prepare, I ran a marathon, hoping to achieve a time that would qualify me for Boston in 2014. I was thrilled to come in under my qualifying time, but that was still no guarantee I'd get into Boston. I had to wait until the fall of 2013, when tens of thousands of runners like me from around the world would sit by their computers, vying for a spot in what promised to be the most historic Boston Marathon ever.

I was ecstatic when my letter of acceptance arrived. I would have a chance to show my love for Boston, for its victims, its runners, and its fans.

I began training in earnest, but the Canadian winter of 2013/2014 was not a training-friendly one. The temperature was torturous and the snow was deep. Because of the icy conditions, I ended up with a hip injury that threatened to dash my hopes.

I tried physio, chiro, massage, rollers, sticks, balls – every way to heal that you can imagine. But when race time came, my wife, Theresa, and I were left to deliberate: Should I bow out? Should we cancel our hotel and our travel plans? Eventually, we decided to go. "Even if I can't run," I thought, "at least I can cheer my fellow runners on and be part of the great crowd. And who knows? Maybe my body would find a way to finish the race, injury and all."

But how would I express my solidarity with Boston?

"I'll take a Canadian flag," I thought. "That way I can declare 'Canada supports you, Boston.'"

No. Support was not a deep enough emotion for what I felt. It was more like love. So I thought, "I'll put a Canadian flag on my shirt that says, 'Canada Loves Boston.'"

No. Not personal enough. It's not just Canada: it's me that loves Boston. So I asked myself, "What is it that I really, authentically feel?"

The answer was immediate: *Brady Loves Boston*. I felt love for the victims, for the city, for the fans, and for the BAA. And so I printed this message on a Canadian flag and put it on the front and back of my shirt.

As every runner knows, when you put your name on your shirt, spectators will call you by name. But I could never have predicted the sort of response my *Brady Loves Boston* message would evoke.

Only minutes into my run I began to hear the cheering. Not general cheering but very personal cheering. One million spectators lined the road from Hopkinton to Boston. Fans were reading my *Brady Loves Boston* shirt and shouting, "Boston loves you right back, Brady." I was shaken. People were looking me right in the eyes and saying, "We love you, too, Brady."

I was amazed and thought, "You don't owe me that. I don't deserve such personal encouragement."

I waved, even blew kisses to the people who cheered for me and said, "Thank you!"

My race was a series of hundreds of mini conversations:

Spectator: "Thanks for coming down here, Canada!"
Brady: "You're so welcome!"
Spectator: "Canada rocks!"
Brady: "Yes, and we love you!"
Spectator: "We love you, Brady."
Brady: "Thank you!"

When a complete stranger reaches out to you and says, with feeling, "We love you" – well, it does something to you.

I confess. I *loved* it. But I began to feel bad for the runners surrounding me – I was the only one being cheered for and encouraged by name. I turned to the man next to me and said, "I'm sorry for this." He laughed and said, "You don't understand: my last name is Brady – I'm not leaving your side!"

By the halfway point, I was in a lot of pain and feared I might have to give up. But the moment I'd think that, someone would shout out, "Boston loves you, Brady!" and my energy surged.

The energy I felt was like nothing I'd ever experienced in my life. It even got me up Heartbreak Hill (actually a series of *four* hills, coinciding with the spot in the race where you "hit the wall"). I never stopped. I ran up all the hills, fueled by the energy of the crowd.

Meanwhile, back home, my kids and friends were tracking my progress online, following my little runner avatar on the BAA website. They were texting Theresa, who was waiting for me at the finish-line, so when I completed the race, she already knew my time. (Most runners wear a watch so they can track their time, but I hadn't bothered to put one on because I was sure I would do so poorly.)

Theresa and I had a tearful reunion, and I said, "Well, it was no great finishing time, but I had the time of my life."

"No, no," she said. "You ran it in 3:48 – only 13 minutes over your qualifying time. You had a fantastic race!"

Engagement vs. Energy

It's energy – not engagement – that fuels high performance. It's *energy* that deploys strengths. It's *energy* that drives execution: translating ideas into action.

From the sun that fuels our existence, to the molecules that hold up your chair, to the electrical flow of neurotransmitters you use to process this paragraph, energy is what makes everything work.

But energy is one of the last things considered in an organization.

When I ask employees and managers what it means to be "engaged," the words on the left-hand column (see below) emerge. As you can see, this common perception differs greatly from the perception of what "energized" means (on the right).

"Engaged"	"Engergized"
Commitment	Passion
Dedication	Drive
Loyalty	Intensity
Care	Devotion
Effort	Enthusiasm
Sacrifice	Resilience
Determination	Vitality

Engagement is essential, foundational, elemental. But by itself it is insufficient. Without energy, it is simply unsustainable. Let's face it: it takes a lot of energy to sustain engagement.

The problem with managing engagement is that employees perceive they are being asked for discretionary effort based on commitment, loyalty, and extra effort. But if they lack energy – if their executive function is shut down – all managers get is dedicated under-performers.

The energized individual, however, brings commitment, loyalty, and effort infused with passion, intensity, and vitality. This makes all the difference in the world. Why? Because *energized employees fuel great customer experiences and better business results.*

Think about it, what do you want your customers saying after they interact with one of your employees? Is this what you want them to say?

She was:
Intelligent
Competent

Helpful

Knowledgeable

Professional

Of course you don't! There is nothing on that list that creates an unforgettable customer experience. Any one of your competitors can evoke these kinds of responses. They are table stakes: the bare minimum required to be in the game. What you want is customers walking away talking about the human magic they've experienced.

She was intuitive.

She connected with me – she really *got* me.

She cared deeply – she found the perfect solution for me.

She was empathetic.

She brought some personality, some sparkle.

These kinds of interactions are not fueled by engagement; they are fueled by energy.

You need people's engagement just to be in the game, but it's energy that creates the customer experience that differentiates you from your competitors.

Managing Energy for Sustainable Results

When managers and employees hear about "discretionary effort," their minds can easily jump to conclusions like, "I need to try harder – think smarter – stay later." But none of these produces sustainable results. Moreover, strengths don't just operate automatically: they are deployed by energy.

For your business to be sustainable, you need a blend of people's discretionary and innovative thought. Intelligent effort. Smart work.

You spend good money hiring people with the right strengths and getting them into the right roles. But engagement by itself just won't cut it. When people are engaged but not energized:

- Customers get the service, but it lacks human magic

- Employees get support, but it feels perfunctory and unimaginative to them
- Problems get solved (for now), but the root causes go unaddressed
- Products go out the door, but they lack brilliance

Energy is what transforms ideas into results. It does so by sparking the dynamic duo of uncommon effort and innovative thinking.

Satisfied, Engaged, Energized

Here is another example to explain how energy transforms ideas into sustainable results.

Meet three customer service employees: Satisfied Shelly, Engaged Elaine, and Energized Eduardo. Here is how each employee might handle a customer complaint.

Satisfied Shelly	*Meets the customer's need.*
Engaged Elaine	*In the process of meeting the need, Elaine uncovers the root cause of the customer's concern; she identifies the inadequacies of the process and shares them with her manager.*
Energized Eduardo	*Does everything Engaged Elaine does but also uses his frontline experience to come up with an innovation that elegantly addresses the issue. Eduardo then uses his influence skills to get the innovation implemented.*

Let's face it: Engaged Elaine is applying effort, but when it comes to sustainable results, effort is necessary but not sufficient. It's *innovative* effort that unlocks and implements breakthrough ideas. The following case provides an intriguing example of this.

Innovative Conversations

Rick Boersma, Juice's innovation practice leader, worked with an 8,000-employee hospital in the U.S. Over the past decade, the hospital strived to instill a culture of innovation among employees focused on "patient-centered care."

Ideally, this culture would be characterized by grassroots innovation, that is, solutions driven by staff working on the frontlines of patient care. This was especially important with nursing employees, who were extremely sensitive to a history of top-down innovations that failed to take their workload into account. A twenty-person innovation pilot group was created, including seven nurses as well as physicians, administrators, and operations personnel.

Rick used one of our tools, Value Chain Analysis, to help this group structure their conversations around innovation, in this case to create a step-by-step map of the patient-intake experience. Rather than asking the broad question, "How can we improve the patient experience?" group members used the tool to break down the experience into manageable parts.

One small step in the intake process included giving patients a box of toiletry items (including toothbrush and toothpaste, moisturizers, mouthwash, shampoo, and a water bottle). This box was placed on their bedside table. The group asked the question, "How could we improve the patient experience around these items?"

Given financial and other restraints, some of the responses were untenable (e.g., "Give them perfume," "Include chocolate"). But one of the ideas generated a lot of energy: "Why not put the items in a gift box?"

Energy Fuels Execution

Innovation without execution is pointless, of course, and the gift box idea could have easily stayed on a Post-it note on the meeting room wall. What happened instead is the real point of the story: *energy fueled a blend of effort and innovation that produced a great patient experience.*

The next morning one of the group, let's call her Frances, came back to her colleagues with a prototype and a plan. She had taken one of the boxes of items and wrapped it in cellophane wrap, adding a bow and a welcome card. The group gathered around her prototype, admired it, and made suggestions.

Ten minutes later, they had a plan:

The wrap and ribbons could come out of their petty-cash budget.

Hospital volunteers could be enlisted for wrapping and to write greeting cards. They could launch the prototype immediately.

The exercise showed how a blend of energy management, skillful conversation, and innovation tools, leaders and employees can co-create cultures that release "small-i" innovations at every level of their organization. If "big-I" innovation is a team of R&D brainiacs sequestered in a brainstorming think tank imagining high-level innovations that will transform the company, "small-i" innovations are the result of every person at every level using simple, structured innovation tools to come up with breakthrough ideas that unlock better results.

The gift-box story shows how great customer experiences bubble up when employees feel energized and have access to simple innovation tools that let them play to their strengths.

Energy Management Question

Would our employees say we're focused on managing energy or engagement?

▶ ▶ ▶ *Manage Energy, Not Engagement*

Deliver Experiences, Not Promises

What's the brain science? Our brain is wired to pursue activities that promise reward regardless of whether the experience of reward is actually delivered.

Where does this show up at work? When leaders promise a great employee experience through engagement programs, employees get involved, but because engagement lacks an energy supply, they soon realize they can't expect to experience the promised benefits.

Why does this matter? A culture that fails to live up to its promise creates cynicism and an unexpected "best-before date" on engagement initiatives. Throughout the business world, employees are seeing employee engagement as a con game.

▶ ▶ ▶

The Promise of Reward

Two young McGill University scientists, James Olds and Peter Milner, set out to research the *avoid* area of the brain – the area that, in experiments with rats, creates such a revolting response that the rats avoid anything associated with that stimulus.

But when inserting the laser-thin electrode into a rat's brain, Olds and Milner missed the avoid center, securing the electrode to another region. In fact, the rat kept returning to the exact location of the cage where it had received the stimulating jolt. Perplexed, Olds and Milner wondered if it *wanted* to be shocked. To test this hypothesis, they gave the rat a tiny shock every time it began to move away from the corner.

Olds and Milner had stumbled on the discovery of an area of the brain that neuroscientists would confirm is associated with *the promise of reward*.

A quick learner, within minutes the rat had followed their beckoning sensations to the opposite corner of the cage. It wasn't long before the scientists were operating the shock-delivery mechanism like the controller of a video game, steering the rat wherever they wanted it to go.

Now they wondered just how powerful this sense of anticipation was, this promise of reward that they were stimulating. To find out, they asked the rat to fast for twenty-four hours (assisting it by removing the food from his cage).

The next day, they placed the rat in a tunnel with food at each end. It smelled the food and raced toward it. But when Olds and Milner gave that tantalizing little jolt en route, the rat screeched to a halt, ignoring the food and refusing to budge from the spot in the tunnel that triggered the promise of reward.

What if the rat were to take the stimulation process into its own hands? To find out, the two scientists rigged up a lever that would trigger the euphoric jolt whenever it was pressed. Once the rat figured out how it worked, it began giving itself a dose of anticipation *every five seconds*. In subsequent tests, other rats also ignored their needs for food and water and manically pressed the lever without ceasing until they collapsed in a heap of exhaustion.

Does this not describe the way many people run from one activity to the next, pursuing the promise of reward while ignoring the things that leave them fulfilled? Is the rat's dilemma our dilemma?

You'd think that we humans, with our superior prefrontal cortex, would never act in such a manner. But a scientist named Robert G. Heath found out we do. He hooked subjects up in Olds and Milner fashion and discovered that, on average, they self-stimulated *40 times per minute*.

Hungry subjects shunned food as they mindlessly mashed the lever. One kept pressing it over 200 times after the power had been turned off. The scientist administering the test finally had to order the subject to stop. It became clear that subjects were not experiencing satisfaction, but only the anticipation of satisfaction; not reward, but only the *promise* of reward.

Is there that much different between us and the rat with its eyes bugging, tongue hanging out, teetering from exhaustion, relentlessly pounding the lever to get that next hit of promise?

Well, consider how many of us sign up for another diet method but end up weighing more. Or buy exercise equipment that goes unused. We're not buying the satisfaction of discipline and a good hard workout. We're buying the promise of a thinner body. We buy the latest smartphone but use only a fraction of its capabilities, because it's not the satisfaction of an ordered life that we purchased but the *promise* of one.

A Powerful Potion

Why do we do this? Because we live in an environment crowded with stimuli that tickle the *promise of reward* region of our brain. The moment that occurs, a payload of dopamine is released and we feel a surge of pleasure. Dopamine (note the "dope" word embedded inside it) is the close cousin of cocaine and morphine. It is a powerful potion that intensifies pleasure and cuts pain. As such, it is insanely addictive.

This dopamine dynamic takes things like cocaine and gambling and produces a hardwiring process in our brain that make these substances and activities insidiously ensnaring. In fact, remove the dopamine response from cocaine and the addiction problem ceases.

We see more innocuous versions of these addictions every day: things like Candy Crush mania, stealth Facebook surfing, and compulsive smartphone use. You've seen the research: if asked to choose between giving up their smartphone for a week or giving up sex, caffeine, or alcohol, a shocking percentage of people would forego the big three in order to keep the smartphone.

I know a senior IT leader who was wooing a high-potential candidate in a job interview. He asked the candidate a question and then, as she began to answer him, snatched up the BlackBerry that was tickling his side and answered the message.

The anticipation that there might be something he needed to know was just too great a temptation. Like the researcher's rat, he pressed the button and got his little jolt – but lost the big reward (the candidate later confided that she would never work for a boss like that).

How the Promise of Reward Short-Circuits Engagement

When it comes to engagement, leaders do things that may be well intended but that clearly have more to do with giving people the anticipation or promise of reward than delivering on the experience of it.

1 We Sell Engagement

Managers meet with their teams to talk about engagement scores from a survey. They start out with the clear intention of listening to their employees' concerns, drawing out what's most important to them and what will help them feel more energized. But before they know it, they are selling the benefits of engagement and telling the team, "I know I can count on you to help turn these scores around." And eyes roll around the table.

Let's be honest: employees don't care about engagement scores, at least not for long. What they do care about is their *lived* experience at work: that they feel they're growing and able to make progress in the things that matter to them and that they have energy left over at the end of the day.

2 We Focus on Rankings and Lists

A CEO talks to his employees about engagement in a town hall meeting. His communications leader and senior team have coached him and carefully crafted his script, telling him, "The employees have to be able to see how this initiative benefits them – how it impacts their experience here, not just how it benefits the company."

But, carried along by the power of the moment, the CEO trumpets, "This year, with all your hard work, our company can make the Top 50 list of best-run companies!"

The leader of communications blanches. The senior team slumps at the table.

Guess what, Mr./Ms. CEO? Just as employees don't care about engagement scores, neither do they care about making the Top 50 list.

3 We Substitute Stuff for Substance

Two chronic themes we see cropping up again and again on engagement surveys are:

- Under-performers are not held accountable
- High performers are not recognized

Leaders gravitate toward elaborate recognition/reward programs and intricate performance management systems that hold the promise of fixing these issues once and for all. But, more often than not, they don't. Instead, inordinate amounts of money are spent on recognition systems that create a sense of democratic recognition. High performers find this insulting because, to them, "democratic recognition" means low performers get recognized the same way they do. So the recognition program gets dubbed with the cynical moniker, TYFDYJ (thank you for doing your job).

Meanwhile, performance management systems are intricate and thorough, so much so that the multitude of subjects that must be discussed and computer screens that must be populated make the process impersonal and onerous. Employees tell us their performance appraisals just never get done – or that, if they do, it seems the focus is on completing the appraisal not developing the employee. The process that promised to fix performance has short-circuited it.

4 We Promise to Fix the Issue for Employees

One of the most common responses by managers to an engagement survey is to look at the low scores and take the monkey on their own back, working even harder to fix issues like communication, work-life balance, and recognition.

Managers who promise to do these things are well meaning and conscientious, but the approach is simply unsustainable. Managers can't fix these issues without partnership from their employees.

5 We Create Aggressive but Unrealistic Strategies

After survey results come out, managers march with their HR partners to create aggressive plans of how they will fix the scores. These strategies are robust but impossible to deliver on, especially with an exhausted workforce that has little capacity.

6 We Pressure Employees

At Juice, we've had to address situations where leaders baldly ask employees to give "fives" when they fill out their engagement surveys: "I know you guys are a great team. I know you're completely engaged. Let's give those fives that show everybody the good work we've done."

The promise of a higher score tickled the brain of these leaders, making them act outside their usual sane and sober behavior.

Let me ask you a question: Do you really want to be managing people's engagement? As a leader, is it really the best use of your time to cheerlead, ask for more loyalty and commitment, plead with employees to say good things about your company?

When you do this, not only are time, money, and energy wasted but your efforts produce unfulfillment, cynicism, and other forms of residue that have to be cleaned up, requiring even further thought, time, money, and energy.

Delivering Experiences

Let's face it: this can all be so much better. In the pages that follow, you'll see how understanding the brain charts a clear course for you through the complexity of employee engagement. For now, let's look at an intriguing example from the world of retirement saving that has something to teach us all about how to deliver an experience of reward, not just the promise of one.

Question: Who's bagging your groceries? A teenager, or a sixty-nine-year-old who had to come back to work to make ends meet? Financial analysts tell us we are facing a retirement crisis. Large numbers of boomers can't afford to retire or have been forced out of retirement due to lack of savings.

How did we get here? Whether because of their shortsightedness, inertia, confusion, or short-term gratification, people simply failed to save enough. But two men are helping to prevent this retirement crisis for the next generation with a simple solution: getting employees to bump up their future savings by having them automatically deposit the extra they get in pay raises. Their current take-home pay remains the same, but future savings steadily balloon.

According to information on knowledge.alliantz.com, "The Save More Tomorrow program developed by Benartzi and fellow behavioral finance expert Richard Thaler uses these solutions and has been taken up by more than half of the large employers in the United States. In its first implementation, the program helped boost the average participant's 401(k) pension plan savings rate from 3.5 percent to 13.6 percent in just 3.5 years."

How does this work? The promise of future reward prompts

a commitment that can be realized because of a simple account-ability mechanism that guarantees downstream results.

- I want a secure future...
- So I commit to invest my next pay raise...
- The system automatically takes that portion and invests it for me

Great coaches and managers do this all the time. They help people get in touch with a strong sense of promise/possibility, ask them for a commitment, and then ensure that an action plan is put in place to increase the odds of follow-through.

Throughout this book, you'll meet leaders and managers who have made the critical shift from the promise of reward to delivery of reward. They have moved beyond non-rewarding engagement activities to helping people experience the reward of an energized environment: an environment where people flourish in the work that matters to them. Instead of delivering promises, they're delivering experiences.

What Matters Most? The Employee Experience

When we ask leaders what matters most to them, they give us minor variations on the same three elements: "We want to create great numbers, a great customer experience, and a great employee experience."

"Fair enough," we say. "Everybody wants those three things. But is there one of them that drives the other two?"

What do you think they say? In fact, what would *you* say?

Most leaders reply, "The employee experience. It's the one that drives the other two."

And they're right. Substantive research by James Heskett et al. shows that if you create a great employee experience, it spills over into both the numbers and the customer experience.

Then we ask, "What if you landed in a rough patch for a couple of months and had to sacrifice one of the three? Which one would you pick?"

Leaders tell us, "We would never sacrifice the numbers, nor would we sacrifice the customer experience. If we had to temporarily give one up, we'd have to pick the employee experience."

This brings us to the crux of why employee engagement is not working. The employee experience is what drives sustainable business results, yet it's the first thing to be cut in moments of difficulty.

The problem is that these rough patches don't present themselves as clearly defined, organization-wide, multi-month rallying crises. In today's work environment, they show up day-in and day-out, month-in and month-out. Let's face it, they make up the fabric of today's work environment:

Are sales soft? "We're going to have to cut vacations."

Is the workload overwhelming? "Time to jettison the recognition program."

Is the technology rollout behind schedule? "Order pizza, because IT is going to be pulling some all-nighters."

Meanwhile, in today's work environment, the employee experience is all about having to do more with less. And what is that, really? Mark Royal and Tom Agnew do a fine job of unpacking "More With Less" in their book *The Enemy of Engagement*. They say that, for many companies,

> it means continually raising the bar on goals and expectations for individuals and departments across the organization,

while at the same time spending less money. That is, it typically involves trying to motivate employees to work harder. And for employees, it means that management is going to want you to work more hours or accomplish more in the same amount of time. And because management is watching costs, you will likely not be given what you need to do the job well. It's frustrating, especially to employees who are engaged, loyal to the organization, and want to achieve excellent results.

The assumption that leaders need to get employees to be more engaged – particularly under such conditions – is an extremely negative notion infused with hidden judgments of those very employees:

- "You're not working hard enough"
- "You're not as committed as you should be"
- "You could be more loyal to this organization"

No wonder employees have begun to see engagement as a con game.

Hard work, commitment, and loyalty to one's company are the natural byproduct of one thing: energy.

Let's remember the message of chapter 1: people can be engaged but not energized. And that's a direct result of leaders not closing the gap between their intentions and what employees actually experience.

This may shock you, but it doesn't matter whether you respect your employees. It doesn't matter whether your employees create value for your customers. It doesn't even matter whether your employees do meaningful work. What really matters is that they *experience* – they *feel* – that you respect them; that they *experience* that they're creating value for your customers; that they *experience* that they're doing meaningful work.

The Box Plop

Imagine you're a leader of a large hospital and you learn from your engagement survey that the nurses feel they lack the basic equipment required to do their jobs. They are constantly running to the other end of the unit to get blood pressure cuffs that should have been at their fingertips.

As you and your fellow leaders look at the results of the survey, you look for the low-hanging fruit, actions that can be quickly knocked off to demonstrate that you've listened and are taking action. You get excited when you see the comments about the blood pressure cuffs.

"Let's buy a box of blood pressure cuffs," one of you says. "When it comes in, one of us will put it on the charge nurse's desk during the night shift. When she arrives for her shift, it will be waiting for her. People will love it."

When we hear these kinds of tactics, we say, "Please don't do this. Plopping down the box on the charge nurse's desk will probably not create the type of experience you're looking to create. Your response could easily be perceived as, 'OK, OK, we heard you already. Take your damn blood pressure cuffs!'"

Senior leaders get the point and ask us, "So what should we do?"

"Meet with the charge nurse and her team and have a conversation with them. The administrator of that area and the CEO can bring the box of blood pressure cuffs and say,

Thanks for your comments on the engagement survey, guys. The blood pressure cuffs were a big miss on our part. We want you to be able to do your job with as little human wear-and-tear as possible.

We know this box of cuffs isn't the big solution to the equipping issue, but it's a sign we're listening. We may not have capacity to do everything or provide everything that could be provided, but if we engage in the conversation together, we'll figure out a way to manage our resources in a way that makes sense for everybody.

It would be so easy for you and your fellow leaders to get swept up in the promise of reward ("They're gonna love it when they see this box of cuffs!") and not deliver the experience of the reward ("You spoke. We listened. We acknowledge that our miss made your job frustrating. Let's begin a better dialogue").

Heed the hard-won wisdom of the paw-scarred rat that chased the *promise* and never got the *experience*. If sustainable results are what matter to you, it's time to shift from activities that promise the reward to those that allow people to experience the reward. That will make all the difference in the world.

▶ CASE STORY
The Engagement Scores That Took Care of Themselves

We were privileged to help Don Ford and the Central East Community Care Access Centre, part of a health-care initiative of the province of Ontario, to do one simple thing: create an employee experience where it *feels good to work and it's easier to get things done.*

We focused managers and leaders on one simple activity: the energizing conversation. All our efforts went to equipping them to partner with employees to remove interference and release energy.

The employee experience at Central East has since significantly shifted for the better, but it's also intriguing to see what happened to the engagement scores.

When we began the journey with CECCAC in 2010, they were clearly disappointed with their scores, which were close to the bottom of the heap, twelfth out of fourteen.

Three years later, they had risen to a very different position: second from the top. The secret? Don and his senior team focused on the employee experience – and the scores took care of themselves. I'm simply stating what you already know intuitively: the scores are just a lagging indicator of the lived experience of your employees.

Now think back to the introduction to this book and the deliberations of senior team members, each giving their reasons for embarking on an engagement initiative. The motive of Don's team stands in sharp contrast to theirs. What was conspicuously missing from that conversation? *The employee experience.*

Energy Management Question

Does our employee experience deliver on the promises we make on our website? How do we know?

▶ ▶ ▶ Deliver Experiences, Not Promises

3

Target Emotion, Not Logic

What's the brain science? Our limbic system – the emotional center of our brain – defines what we experience as reality. Mirror neurons make all of this possible. They detect and reflect others' actions, emotions, and even intentions.

How does this show up at work? *People experience only what they feel.* It's the emotional brain that defines people's lived experience at work. The emotional brain provides us with an uncanny knack for recognizing care, support, and respect. It also enables us to recognize when care, support, and respect are not present but are simply being declared by the other person.

Why does this matter? Employee engagement gets mired down when leaders fail to grasp that emotional engagement, not rational engagement, is what defines the employee experience and unlocks their discretionary effort. Targeting emotion closes the gap between what you intend as a leader or manager and what your employees actually feel.

▶ ▶ ▶

Four Times the Effort

What causes employees to go above and beyond the call of duty (ABCD) – to go the extra mile, to give what's called their *discretionary effort*?

It's not giving them a sexy laptop. That's just expected. It's not awing them with your intelligence. Sorry. It's not even impressing them with your dedication as they see you come in earlier, stay later, and work more weekends than anyone else. Employees look at all that and say to themselves, "Knock yourself out...I'm going home to my kids."

There are two types of engagement that unlock discretionary effort: rational engagement and emotional engagement. The Corporate Leadership Council (CLC) did a study of 50,000 employees worldwide and found that emotional engagement releases four times the discretionary effort. Welcome to the feelings economy, where *what employees feel* matters most. Consider the differences between rational and emotional engagement.

Rational Engagement	Emotional Engagement
Engaging the minds of your employees	Engaging the hearts of your employees
• "I understand the expectations my manager has of me and what I need to do to be successful"	• "My manager listens to me in a way that makes me feel valued and respected"
• "I understand the big picture and how I contribute to it"	• "My leaders take interest in me as a person"
• "I understand our global goals and how they chunk down into my day-to-day objectives"	• "I am proud of the purpose of my company"

Every pay period, your employees receive two paychecks: a fiscal one and an emotional one. The fiscal paycheck releases no discretionary effort; it's just expected. But the emotional paycheck causes people to go ABCD.

Why does emotion trump cognition in this way?

Science Explains This

As reported by the writer Gavin de Becker, brain research reveals that your limbic system – the emotional center of your brain – *defines what you experience as reality.*

This inner sensor works in the following way. All data entering your prefrontal cortex (the logical, decision-making center of the brain) are first filtered through your limbic system. In short, as research on the amygdala and prefrontal cortex show, you feel before you think.

The limbic system then assigns meaning to the incoming stimulus (by cross-referencing millions of data points from your emotional history) and registers it as an emotion. It's this emotion that instructs you about what is real, what is true. In other words, if you don't *feel* you're valued, all the assertions, declarations, and assurances in the world won't make it true for you.

All of this happens fast – faster than conscious thought. A recent NYU study by Jonathan Freeman et al. showed that it takes only 50 milliseconds, or a 20th of a second, for the amygdala, an almond-shaped structure in the frontal portion of the brain, to judge whether a face is trustworthy or not. (This amount of time is insufficient for a conscious assessment.)

These lightning-fast judgments have enabled us to thrive and survive. When the stakes are high, whether we're faced with an opportunity (food or sex) or a threat (predation, severe weather, or combat), our emotional sensors deliver up an instant solution, ensuring we don't forfeit the opportunity or escape the danger by wasting precious seconds pondering things through. We're here because our ancestors' emotional brains knew what/who to trust long before their rational brains did.

All of this means your employees' first response to you is an emotional one. At the outset of an interaction with you, they are internally assessing, "Do I feel put down? Respected? Listened to? Patronized?" If they don't *feel* valued by you as their boss, all the assertions, declarations, and assurances in the world can't make it true for them.

And it's not just the primal wiring of approach/avoid that forms our judgments. Our brain comes equipped with a sophisticated set of mirror neurons that detect the presence or absence of another's emotions.

These mirror neurons recognize when care, support, and respect are present in the person we're interacting with. They also recognize when care is being declared but disdain, contempt, and boredom are being telegraphed. It's sobering for us leaders to realize that our employees can sense whether we're authentic or not.

And emotions not only start the assessment process; they end it. When your interaction with someone is over, you may not remember what the other person said – but you will definitely remember how he or she made you feel. As Hadley and MacKay have shown, our feelings are the first and last things we remember about any experience, with rational thinking simply sandwiched between.

If you don't step in to your employees' world and find out what matters most (WMM) to them, what you intend will fall into the gap between your world and theirs, and it won't be felt. And if your recognition, inspiration, support, or respect is not felt, the employee's emotional brain says it simply is not real.

I had a woman tell me, "Our company had a banner year last year, and our manager wanted to recognize our team. So he took us all out to a hockey game."

I asked her how that went.

"Our department is made up mostly of women," she said. "I know some women like hockey, but *it means nothing to our group*. If he had taken us out to dinner and a play, we would have felt completely recognized."

Too bad. This manager had invested time and (lots of) money trying to recognize his employees, but they still didn't feel recognized.

Let's be clear: no matter how noble your intentions are, recognition is not truly recognition unless it's felt. Remember, in the feelings economy, it's what is felt that counts.

How often does this happen in organizations? I see it far too often. What a waste when so much money, time, and effort falls into the gap between what is *intended* and what is *felt*. It's bad enough that these resources fail to produce the desired effect. What's worse is that sometimes the effort to inspire, support, or recognize actually produces the *opposite* of what was intended. Employees feel offended, abandoned, or even betrayed.

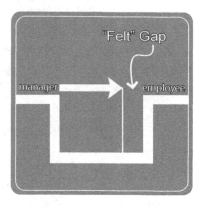

It's not difficult to identify the source of this dynamic. When it comes to selecting leaders, organizations prize reason over emotion. But research is unveiling the fact that it may be time to change the game.

"Soft" Feelings Drive Hard Results

Marcus Buckingham and Curt Coffman make the strong suggestion in their book *First, Break All the Rules* that "it would be [most] efficient to identify the few emotions you want your employees to feel and then to hold your managers accountable for creating these emotions."

These emotions then become the outcomes a manager is primarily responsible for.

Perhaps understanding what brain science says about emotions will help us put to bed the myth that you have to be calculating and cerebral to be a great leader.

Kouzes and Posner artfully address this myth in their book *Encouraging the Heart*. They cite a study conducted by the Center for Creative Leadership (CCL) focusing on the factors that account for a manager's success. The CCL made the surprising discovery that there was only one factor that significantly differentiated top-quartile managers from bottom-quartile managers: higher scores on *affection*, both expressed and wanted.

Contrary to the myth of the cold-hearted boss who cares very little about people's feelings, the highest-performing managers show more warmth and fondness toward others. They get closer to people and are significantly more open in sharing thoughts and feelings than their lower-performing counterparts.

Now, these managers were not without their rational sides. In fact, on another measure administered by CCL they all scored high on "thinking" and on their need to have power and influence over others. It's just that these factors didn't explain why managers were higher performers.

Want to manage energy and create a great employee experience? The research tells you to target emotions, show warmth, connect, put words to your feelings.

These are not techniques or skills but comprise *a way of being* that will help you close the gap between what you intend and what employees feel.

Sleuthing Out What Matters Most

People's felt needs aren't typically tattooed on their foreheads. So how can you sleuth out what matters most (WMM) to them in any given situation?

There is one simple but brilliant activity that great managers do that lets them know what matters most to their employees: *they ask*. Let's say the employee is talking about flextime. The manager asks, "That's important to different people for different reasons. What is it about flextime that matters most to you?"

This simple, straightforward approach guides managers to the employee's picture of felt support. Once managers get that picture, they check their understanding with another straightforward question: "So if you and I are able to _____, would you find that supportive?"

When managers and leaders learn how to step in to their employees' worlds, they can get skillful at discovering what it takes to close the gap between what they intend and what the employee feels. But your employees' worlds can be as different from yours as Beijing is from Boston.

Stepping in to Other People's Worlds

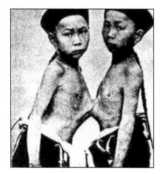

Have you ever heard of the Bunker conjoined twins? Their names were Chang and Eng Bunker. Born in what was Siam and is now Thailand, they were discovered by Barnum & Bailey and toured with that circus.

Take a moment to step in to their world and think of what it would be like to have been the Bunkers.

When we ask groups of managers to do this exercise, they tell us things like:

- I'd get really tired of being a spectacle
- Everything would take more effort
- I might feel somewhat exploited by the circus
- I'd wonder what kind of a future I could hope for

Then we ask participants, "What if we told you that Chang and Eng used their savings from the circus to buy a plantation in the Carolinas? Does that change your picture?"

This often causes participants to give their heads a shake, but the next bit of information we give them is even more astounding: we show a picture of them with their wives and families. Chang and Eng fell in love with two sisters, ended up marrying them and having large families. They alternated their visits with each wife: three days with one, then three days with the other. The picture suggests that they figured out how to make things work (on more levels than one).

When we ask participants, "How did you do at stepping in to the world of Chang and Eng Bunker?" they usually tell us, "Not so great. We were still stuck in our own worlds."

This highlights the problem with entry-level empathy: putting yourself in someone else's shoes.

Why is that a problem? *Because it's still you in their shoes*: you, with all your assumptions, pre-formed conclusions, and judgments. Even the language we use betrays this mindset: "If I were in your shoes, I would _____."

Whatever you insert into that blank will be yours, projected onto their world: *your* values, *your* beliefs, *your* autobiographical prescription of what they should do.

This is a surefire recipe for *intended*, not felt support.

Do something different. Step in to your employees' worlds and put on their eyes. See the world as they see it. This is a much more personalized type of empathy. When you get behind their eyes and look out at the world, you'll see and feel the situation the way they see and feel it. The action you take next will likely hit the bull's eye of the felt need – and create a poignant employee experience.

When you discover what matters most to your employees, you can invest your time intelligently, partnering with them to locate *their source of energy*. As you'll see in the following story, the results of doing this can be almost magical.

▶ CASE STORY
"1.5 Employees in One Body"

Alan, an IT manager, inherited a new team, and with it, a challenging employee.

Tina was a business analyst who had spent the previous two years experiencing several life setbacks. She struggled with absenteeism, stress-induced illness, and a general sense of dissatisfaction in her work situation. As a result, she had begun to wear the scarlet letter, branded as an employee who might need to be exited from the organization.

But as Alan observed Tina in her role, he discovered that she was very smart and still had a lot to contribute – if only he could figure out what was standing in the way of her success.

As Alan tells the story, even though he clearly intended to support Tina, he failed to step in to her world and find out what mattered most to her. Instead, he made an assumption that shifted him from partnering with her to parenting her:

"I started mother-henning her and said, 'Tina, I know there's lots going on in your life. We've got to find a way to take some stuff off your plate.' But our conversations were unfulfilling for both of us, and her performance failed to improve."

Alan was baffled. If it wasn't a workload issue, how could he get Tina back to being a productive employee?

In the interim, he took part in a pilot of our engagement process and learned how to do a simple Energy Check conversation: stepping in to someone's world and drawing out *what matters most* – the driving need that's most important to the employee in the midst of her situation.

Alan began meeting with Tina to do these conversations. It was only when he directly asked her, "What matters most to you? How can we work together so you can do your best work here?" that he discovered what felt support was from her perspective. To his surprise, he learned that Tina wanted *more challenges, not fewer.*

"With so much turmoil in her personal life, Tina felt the one place where she could build and demonstrate her value and significance was at work," says Alan. "This was a profound awakening to me – and it told us that leaders need to consider the distinction between intending their support and actually ensuring that their support is felt."

As a result, Tina began to feel completely supported by Alan, and consequently began knocking things out of the park, delivering great results on every new assignment and responsibility he sent her way.

Today, Alan refers to Tina as "1.5 employees wrapped up in one body." Our engagement approach gave him and others in his organization the tools and skills to help them and their employees describe how they felt about things – particularly things getting in the way of their energy levels and engagement. "Before, we didn't know how to put this stuff into words," he says. "Now we have the language we need."

Tina's pre- and post-Energy Check could not have been more different. Her letter to us says it best:

I had allowed things to spiral out of control to the point where I was simply looking for a way out. It's no small feat to make a complete 180-degree change to a career that was on a downturn like mine was, but now I know it's possible. I am achieving things I wouldn't have even considered possible this time last year.

I feel confident, energized, and challenged every day. I arrive early to work and often stay late without even realizing it. I am taking on new challenges just as fast as my manager can throw them my way. I love coming to work and no longer struggle with illnesses that have plagued me in the past. I finally feel my career is headed in a positive direction.

Epilogue: Tina was later nominated for a Bravo Award for high performers, and at the time of this writing, she has been promoted to a managerial position. If Alan had not learned how to show felt support, this superstar could have been lost to the organization.

As a leader, closing the gap between your intention to support and someone feeling your support changes everything that matters: the employee experience, the customer experience, and the numbers.

Energy Management Question

What can we do to ensure that our support comes across as felt, not just intended?

▶ ▶ ▶ *Target Emotion, Not Logic*

4

Trust Conversations, Not Surveys

What's the brain science? Quality conversations release energy and high-performance hormones in the brain.

Where does this show up at work? Leaders don't recognize that conversation releases energy so they trust machinery (the survey, the strategies, the town halls, the newsletter, and the departmental action plans) to generate employee engagement.

But that's not how human energy is generated. It's generated by conversation, the millennia-old medium recognized by the human brain. Energy flows idiosyncratically – through relationships – and when it comes to engaging employees, there is one relationship that trumps all others in its ability to generate (or deplete) energy: that of the manager with an employee.

Why does this matter? A good conversation not only generates energy in employees, it also saves managers time and mind-space. Why?

Most leaders take a retroactive approach to engagement, using the survey (after a year has gone by) to discover what's going wrong and then reacting to the low scores. But these repair conversations consume 10X the mental energy of prevent conversations.

When issues and concerns fester and simmer, they turn into "crucial," "fierce," or "difficult" conversations. In essence, we've relegated conversation to the episodic.

We've been so focused on calamity-based conversations that we've ignored the larger fabric of everyday life. Every conversation matters, and when managers learn to have energizing ones, they can pre-empt and avert issues long before they become chaos-infused.

By the time issues become calamity-based, there's so much residue associated with them that multiples of energy and mind-space are sucked from everyone in the organization. This robs managers of time and plunks them into a limbic stew of reactivity and mental anguish.

Could you take a survey sabbatical? With all the money and time you'll save, you can focus your efforts on short, frequent energy conversations. These conversations release brain-friendly hormones, unlock high performance, draw out real-time intelligence, and address issues before they become episodic. Besides energizing employees, they lift your burden as a manager.

▶ ▶ ▶

The Chemistry of Conversation

In the previous chapter we saw that our emotional brain tells us what is true, that it knows whether we can trust someone long before our rational brain does. So when someone says, "I value you," our emotional brain, not our rational one, senses whether that "value" is genuine.

But the emotional centers of our brains are more than reality-sensors and B.S. detectors. They are electrochemical conduits through which we transmit emotions to one another.

What does this mean to you as a leader or manager? When the strong emotional currents of value, respect, and care are truly present within you, you telegraph them to your employees. This primes positive emotions in them that release a flow of high-performance hormones, or what Christopher Bergland has called "the neurochemicals of happiness."

These hormones unlock more than effort. They unlock intelligence, too. They boost the brain's processing power because they form an energy cocktail of connection, calmness, concentration, creativity, and curiosity – basically Miracle-Gro for the brain.

The generation of human-to-human energy is neither ethereal nor mysterious. It can be mapped inside the brain. Energy is generated electrochemically as hormones are released through quality conversations. Let's explore how this happens, looking at three things: our open-loop limbic system, "mirror neurons," and the individual *hormones* and their effects.

1 Our Open-Loop System

You never have to be concerned about your blood getting mixed with other people's by being close to them. That's because your circulatory system is a closed-loop system. In contrast, your limbic system, the emotional center of your brain, is an open-loop system. Your emotions can be affected – even regulated – by others. Other people's tears, smiles, or expressions of disgust can trigger an involuntary sympathetic reaction in you.

In *Primal Leadership*, Daniel Goleman et al. cite studies in which scientists measure the heart rate of two people as they have a good conversation. At the beginning of the conversation, their bodies function at different rhythms, but fifteen minutes later "their physiological profiles look remarkably similar – a phenomenon called *mirroring.*"

He adds that "scientists describe [the limbic loop] as 'interpersonal limbic regulation,' whereby one person transmits signals that can alter hormone levels, cardiovascular function, sleep rhythms, and even immune function inside the body of another...The open-loop design of the limbic system means that other people can change our very physiology."

Put us together in face-to-face conversations and we regulate each other's emotions. You've probably experienced this yourself. One team member's strong, buoyant mood affects one person after another until the whole team feels upbeat. Another member's critical, negative mood has the opposite effect.

Goleman goes on to say: "This circuitry also attunes our own biology to the dominant range of feelings of the person we are with, so that our emotional states tend to converge. One term scientists use for this neural attunement is limbic resonance, 'a symphony of mutual exchange and internal adaptation' whereby two people harmonize their emotional states."

As a leader, you are always sending out emotional wavelengths. Those wavelengths travel effortlessly through the open-loop system of your employees and begin to resonate inside them. Let's turn now to the specific neurons in our brains responsible for this mirroring process.

2 The Mirror Neurons Inside Your Brain

I once watched a cyclist lose control and crash into a lamppost. The man was a stranger – I had no connection with him whatsoever – yet a wave of revulsion rushed up my spine, and I cringed as though in pain.

This is a universal phenomenon that has mystified psychologists for decades. What's behind these sympathetic responses is why books and movies captivate us so. We find ourselves deep inside the experience of the main characters. We groan, are startled, and even cry as we mirror their emotions.

An exploration of this phenomenon emerged in the late 1980s at the University of Parma. There, an Italian research team led by Giacomo Rizzolatti discovered a peculiar type of neuron in macaque monkeys: cells that fired when the monkeys took an action (like holding a peanut) and when they simply watched the action performed (when a man held a peanut). Further discoveries caused the scientists to dub these *mirror neurons*. brain cells that fire when an action is performed but also when the action is simply observed.

Scientists were eager to determine if this same phenomenon held true for humans. But it was hard to find subjects willing to take part in the research. For some reason having electrodes attached to their brains seemed a bit invasive to potential participants.

Functional magnetic resonance imaging (fMRI) changed everything. It allowed researchers to offer a non-invasive approach, which gave them access to broad groups of subjects. Many studies have now been conducted to determine whether mirror neurons function the same way in humans as they do in primates. For example, William Hutchison and his colleagues at the University of Toronto showed that the same mirror neurons that fire when a subject's hand is pricked with a needle also fire when the subject watches someone else's hand being pricked.

Another study, by Keysers et al., documented in the journal *Neuron*, showed neurons firing when subjects were lightly touched on the leg with a feather duster. The same neurons fired when they simply watched others being touched in the same spot.

As reported by Wicker et al. in another issue of *Neuron*, our mirror neurons fire when we witness another's actions, but do they also fire when we perceive people's emotions? Through a study of an emotion that's hard to mistake, disgust, researchers discovered that they do. The researchers had participants inhale the smell of putrid, rotten butter. Brain imaging (fMRI) revealed that a specific region in the olfactory (smell) area of the brain lit up when they grimaced in disgust. Then they showed participants a video of an actor's face as it twisted into an expression of disgust. The same area of the brain lit up again.

Small wonder that a jolt of pain shot up my spine as I watched the cyclist hit the lamppost. It's primal empathy in action: we experience what others experience. People's actions and their emotions are mirrored inside us because of the intelligence of these unique neurons.

This sophisticated mirror system gives us a read-out of the emotion that is being *projected* by others. As Rizzolatti said, "Mirror neurons are how we recognize an emotion in others neurally."

So far we've seen that mirror neurons enable us to read people's actions and also their emotions. But through subsequent research scientists discovered that mirror neurons can also decipher the intent behind people's actions. In short, we can anticipate what someone is going to do because our mirror neurons are mind-readers.

This makes perfect sense from a survival point of view. The lightning-fast assessment powers of mirror neurons enabled clans ten thousand years ago to survive when another clan suddenly appeared in front of them, spears in hand. Today, they allow us to thrive, to create rich, human moments with others as we read, recognize, and experience their emotions at an unconscious, neural level.

Rizolatti's research documents that our mirror neurons facilitate the open-loop system through which we imprint one another with emotions. They provide a portal of rich connection, a wide-open conduit of unconscious human interaction. And this emotional superhighway enables us to trigger a flow of high-performance hormones in one another that can draw out people's best thinking.

Earlier I said that innovation is impossible without a rich supply of energy to the brain's executive function. But the energy I'm talking about is not some generic current unrecognized by the brain. Innovation is fueled by three hormones as intimately known to your brain as the food (glucose) it snacks on every day.

3 A Trio of Hormones

If someone told you there was a pill that had absolutely no side effects and was proven to boost creativity, drive goal achievement, fuel learning, enhance pleasure, build trust, deepen relationships, regenerate heart cells, and reduce stress, tension, and fear, you'd say, "Sounds too good to be true," and then you'd take the pill. Faithfully. Every morning.

Does such a wonder drug exist? Yes – but not in pill form.

In fact, every one of the benefits above are available to us through three hormones that boost brain performance: dopamine, oxytocin, and serotonin. (For the purposes of brevity, I will refer to these as hormones. The truth is, they are both hormones and neurotransmitters: hormones when acting in the blood stream; neurotransmitters when acting in the synapses.)

1 Dopamine and the Seeking Drive

Do you know someone who seems frozen, disengaged, uninterested, listless, and unmotivated? It is likely that he or she has a dopamine deficit. Why? Dopamine primes what's called the seeking drive – nature's way of ensuring we get to eat.

Having mapped emotional systems of the brain for several decades, Professor Jaak Panksepp of Washington State University says, "Seeking is the 'granddaddy' of all emotional systems." As Emily Yoffe writes in *Slate* regarding Panksepp's views, "It is the mammalian motivational engine that…gets us out of the bed, or den, or hole to venture forth into the world."

As described in his book *Affective Neuroscience*, Panksepp has explored many words to capture the essence of this drive, including *expectancy*, *interest*, and *anticipation*. But none captures the urgent sense of eagerness and reward-orientation involved like the word *seeking*.

Panksepp says the juice that powers up the seeking drive is dopamine. He calls it the primary contributor "to our feelings of engagement and excitement as we seek the material resources needed for bodily survival, and also when we pursue the cognitive interests that bring positive existential meanings into our lives." Block dopamine, he says, and "human aspirations remain frozen…" But when dopamine is flowing, "a person feels as if he or she can do anything."

It's easy to see, then, why dopamine is such a high-performance hormone: it triggers the seeking drive that turns over every stone until it uncovers the missing innovation. A group of sharp thinkers in a room without dopamine will generate pedestrian innovations at best. Add dopamine and you boost these people with:

1 Creativity

2 Concentration

3 Cognition

4 Goal-direction

5 Reward-orientation

6 Meaning-making

7 Pleasure

These elements create a rich ecosystem that spawns not just evolutionary but also revolutionary innovations.

And speaking of revolutionary innovations, it turns out that dopamine provides another element of an even higher order, a prescient type of *intuition* from which transformational ideas are sparked. A fascinating stock market study by Dr. Read Montague and his colleagues Kenneth Kishida et al. has revealed that dopamine enables people to *intuit* and *anticipate* future patterns.

These researchers wanted to discover what happened to dopamine levels *just before* stock market trades were made. They determined that "the slope of the dopamine signal over a period five seconds prior to a market price update correlated with subsequent market returns, demonstrating that it is a significant predictor of future market activity."

They tested "the capacity of this prediction" by constructing "a trading model based on the fluctuations in the dopamine signal leading up to the market price updates, which invested 100 percent (all in) when the dopamine slope was positive and 0 percent (all out) when the slope was negative. Over the 5 markets played, this trader model earned 202 points (a gain of 175 percent), more than two times the amount earned by the subject's expressed behavior."

2 Oxytocin: The Cuddle Chemical

But dopamine doesn't work as a lone ranger, and intuition, although powerful, isn't sufficient for optimal brain function. As documented by Edward Hallowell, oxytocin – the second high-performance hormone – releases another vital element of innovative energy: openness.

Oxytocin is present to some degree in all of us, both men and women, but it is most abundant (and evident) in mothers as they pour care into their infants in the nursing process. That makes perfect sense. Survival dictated that mammals needed a surefire way of managing the perils of live birth and care of offspring. Oxytocin kick-starts the maternal caring instinct in mothers as well as predisposing infants to stay close to them: two keys to the survival of any species.

Oxytocin is a bonding hormone. It prompts trust, connection, rapport, empathy, compassion, and generosity – everything you could ever hope for in a mom (and what every mom wishes for but never really expects from her child). All of these elements have a singular effect when triggered inside employees: openness.

Oxytocin (like dopamine) is a hormone that is proximity-driven. When you are remote from another person, your oxytocin levels go down and you feel less bonded with them. You also feel less empathy for them. That explains why it's so easy to whip yourself into a frenzy and rip someone apart by email.

But come into proximity with a person – especially in a quality face-to-face conversation – and your oxytocin levels become elevated. In-person connection also releases dopamine (more intuition) and serotonin (more calmness).

But surprisingly, the release of oxytocin is not confined to those magical mind-melding moments. It is also pumped out in moments of stress. As Kelly McGonigal says in her excellent TedTalk "How to Make Stress Your Friend," "Oxytocin is a neurohormone that fine-tunes your brain's social instincts. When oxytocin is released in the stress response, it is motivating you to seek support." She adds, "Your biological stress response is nudging you to tell someone how you feel instead of bottling up. When life is difficult, your stress response wants you to be surrounded by people who care about you."

And oxytocin is more than the brilliant coach in your cranium, nudging you back into relationship in times of stress: it's your on-board cardiologist as well. Oxytocin acts as a natural anti-inflammatory, McGonigal says, relaxing your blood vessels, protecting your cardiovascular system, and regenerating your heart cells.

Here's an interesting equation to consider: (Smart) Trust = Prosperity. As Dr. Paul Zak, a pioneer in the field of neuroeconomics, has shown, oxytocin has much to do with trust. Zak wanted to discover why poor countries were poor and rich countries were rich. The answer was trust. "Levels of interpersonal trust were among the strongest predictors economists have ever found to understand why poor countries are poor and don't grow. Rich countries are by and large high-trust countries."

Here's where oxytocin enters the equation. Oxytocin connects us to others and helps us understand them. "Which," as Zak says, "leads to moral behavior. Which leads to trust. Which leads to a greater number of economic transactions at lower transaction costs."

And the beauty of oxytocin is that it doesn't predispose us to blind trust and gullibility. "It modulates approach/withdrawal," acting like a delicate sensor, "helping us to maintain our balance between behavior based on trust and behavior based on wariness and distrust. In this way oxytocin helps us to navigate between the social benefits of openness – which are considerable – and the reasonable caution that we need to avoid being taken for a ride."

Survival was, and is, dependent on learning that some people should not be trusted and others should be. Oxytocin tells which is which. When oxytocin senses the other person is safe, it goes to work to make us more moral, Zak says. "When you turn on oxytocin in the brain, people engage in virtuous and unselfish behavior."

He and his researchers performed a study in which participants were infused with oxytocin (via nasal spray) or a placebo (saline spray). Asked to make an instant decision "on how to split a sum of money with a stranger...those on oxytocin were 80% more generous than those given a placebo."

A simple way to trigger the release of oxytocin in another is to give them a sign of trust. The economics games that Zak uses in his research reveal that "when one person extends himself to another in a trusting way – by, say, giving money – the person being trusted experiences a surge in oxytocin that makes her less likely to hold back and less likely to cheat."

Great managers realize that sending a trust signal causes a good employee to rise to the occasion, demonstrating greater levels of trustworthiness. This, in turn, causes the manager to be even more trusting, and a virtuous loop is created that shifts the burden of responsibility.

You can't turn oxytocin on in yourself, but you can release it in others. According to Zak, simple activities such as face-to-face interactions, healthy physical contact, team-bonding exercises, and increased exposure to others outside the "tribe" can release the flow of oxytocin in others – in your case, your employees.

3 Serotonin: Signal Strength for Your System

Dopamine and oxytocin would have a muted effect in your brain and body without serotonin. Its job is to regulate the signal intensity of other neurotransmitters, and it influences virtually every neuron in your brain and every cell in your body.

In fact, serotonin not only alters behavior but also organizes and integrates behavior throughout your body. Researchers have found that a properly operating serotonergic system is critical to well-being and relating to others.

This system:

- Reduces the sense of fear
- Decreases tension
- Lessens worry
- Cuts stress

And all of this makes people, in a word, *calm*. But take calmness away and any of the behaviors associated with low serotonin levels could show up: anxiety, panic attacks, obsessive-compulsive behaviors, anger, depression. Managing energy becomes a monumental job for leaders whose employees are living with these challenges.

Clearly, being calm (or not) impacts social skills. For example, healthy serotonin levels make people more pro-social, producing an aversion to harming others.

Serotonin selectively influences moral judgment and behavior through effects on harm aversion, according to a study led by Molly J. Crockett.

On the flipside, reduced serotonin levels produce more aggression and impulsivity in decision-making.

Energy Check Conversations

What if your employees could be primed with healthy levels of dopamine, oxytocin, and serotonin? Imagine how this would affect their energy. Visualize your team sitting around the meeting-room table:

- Goal-oriented
- Creativity-charged
- Connection-primed
- Trust-disposed
- Fear-free
- Focus-filled

It's not a stretch to believe that innovative thinking would flow. At Juice Inc. we've discovered that the three performance-energizing hormones we've been talking about are all unlocked by one simple activity: something we call Energy Check conversations.

Organizations are using Energy Check conversations at all the synapses, or interfaces, that matter to them:

- Manager to employee
- Manager to boss
- Employee to employee
- On teams
- With internal customers
- With external customers

Energy Check conversations co-create the conditions that release the flow of dopamine, oxytocin, and serotonin, perfectly addressing the five driving needs of anyone in the workplace:

1 **Belonging** is served by oxytocin, which creates connection, attachment, and bonding.

2 **Security** is served by serotonin, which creates calm and reduces fear, stress, tension, and worry.

3 **Freedom** is served by dopamine, which cuts pain and boosts creativity, and nor-epinephrine (synthesized from dopamine), which, in eustress (good stress) situations produces an ability to sustain vigilant concentration.

4 **Significance** is also served by dopamine, which fuels reward-orientation, goal-direction, and the seeking drive.

5 **Meaning** is served by catecholamines (dopamine, nor-epinephrine, and epinephrine acting in concert), which trigger intense fascination, curiosity, and focus.

Energy Check conversations create an environment that releases renewable energy, an environment that is challenge-healthy, goal-fueled, distraction-free, feedback-rich, and meaning-infused. Through these conversations, managers and leaders:

1 **Find the fit by:**
- Job-crafting
- Chunking up the challenge to create just the right amount of stretch
- Exploring learning and growth aspirations
- Engaging in career conversations
- Initiating and encouraging others to initiate social, team-bonding, and team-building events

2 **Co-create clarity by:**
- Sparking insights with powerful questions that create tension between their current thinking and the thinking demanded of them to create breakthrough results. Insight and possibility release a jolt of electricity inside the brain that can recharge energy in seconds. As David Rock has put it, "When people make deep new connections in their mind, on a physical level, this aha moment releases chemicals in the body to prime it for movement"
- Co-creating clear expectations of one another
- Asking the questions that cultivate decision-making capacity and thinking skills ("What have you thought of so far?")
- Identifying agreed-on goals
- Removing priority creep
- Creating big picture clarity

3 **Support to success by:**
- Jointly identifying with employees the sweet spot of autonomy and support
- Challenging employees to reduce their circle of concern by growing their circle of influence
- Co-determining the felt support they need to do this

4 **Recognize people's value by:**
- Creating a feedback-rich state

- Affirming the behaviors that are working and nudging the ones that aren't
- Frequently acknowledging the employee's contribution to the success of the business

5 Spark inspiration by:
- Making meaning by connecting the employee's why to the why of the organization
- Challenging the employee to drive toward meaningful progress
- Helping the employee to find meaning at work
- Increasing the frequency and quality of face-to-face (or at least voice-to-voice) conversations

All these activities create a blend of both rational and emotional engagement that give people the energy they need to go ABCD (above and beyond the call of duty) and blow away their customers and co-workers with amazing experiences.

▶ CASE STORY
Conversations at Co-operators

Sandy Ram is an HR leader with a vision: he wants energy conversations to be a part of the fabric of how business is done at Co-operators. As you'll see from the results, after a year and a half of championing a successful engagement process, he's well on his way to seeing that vision become a reality.

Co-operators is a Canadian-owned co-operative well known for its community involvement and commitment to sustainability. The enterprise has a common goal: to be where Canadians are, with the products Canadians need, when they need them, and however they wish to buy them.

Attaining that goal requires a highly engaged workforce and Co-operators is no stranger to success in that area. In fact, Aon Hewitt has honored The Co-operators on more than one occasion as among the 50 Best Employers in Canada, most recently in 2013.

In late 2012, Co-operators began to see its employee engagement scores falter. Leaders sought answers to understand why. What they discovered was that leaders and employees were engaged to a certain degree, but exhausted in their roles.

Traditional business and human resources efforts were realizing minimal to modest success but were not sustaining the zest and passion of employees and leaders. Moreover, recent layoffs had only added to their energy depletion.

Consequently, the co-operative embarked on a journey to change the way they did engagement: one that focused not on scores or numbers, but on building the conditions that energize employees.

"You're Invited to the Party"

At a breakfast in Regina on our engagement process, we met several members of the Co-operators' HR team. When we got to the point where we said, "People can be engaged but not energized," the group started to get excited. Afterwards, I asked why.

"Our VP of HR and our President just did a series of town halls. Leaders across the country told them the same thing: 'we are engaged, but we are exhausted.'"

Clearly the time had come to learn how to manage energy rather than engagement. And it had to be a simple system that would be easy to implement and would not deplete employees' or leaders' energy levels any further.

"We didn't want our leaders to look at this initiative as additional work, or that we were adding another layer to their current demands," says Ram. "We wanted to bring ease to what they were doing."

At Juice, we believe engagement should not be done to employees or even for employees but with employees. That's the essence of partnering: not *to* or *for* but *with*. For that reason, we've learned to take a pull rather than a push approach to engagement initiatives.

We decided we would simply meet with managers and leaders and invite them to the party. Those who wanted to come and were ready to invest the time and energy would get to be part of the process. We wouldn't waste our time and energy on those who weren't ready. We knew that when the uninterested heard the right kind of results stories emerging from the pilot group, readiness and appetite would take care of themselves.

So we launched a three-month pilot on managing energy. A dozen managers came to the party, each one representing a different function across the organization. Several of them brought significant engagement issues and flagging survey scores to share.

We put a robust system in place (see more on the process at juice-check.juiceinc.com) to ensure these managers would be successful doing one thing: the energy conversation. Called Energy Checks, they happen anywhere partnering needs to happen. Managers in this engagement process learned how to do them with:

- An employee, one-on-one
- Their team
- Their internal customers
- Their external customers
- Their boss

Co-operators decided to focus on two of these: the one-on-one Energy Check and the team Energy Check. These simple, quick check-ins helped to pinpoint what was affecting employees' day-to-day energy levels. They allowed managers and employees to co-create the conditions in which they and their teams could naturally go the extra mile in their jobs.

At the end of three months, the organization compared participants' pre-pilot and post-pilot engagement scores as reported in the company-wide engagement survey. The majority reported an increase in engagement, some with statistically significant jumps of as high as 24 to 33 percent. This created excitement.

However, says Ram, "The scores are not our focal point. Our objective is to release energy and remove interference that's getting in the way of that energy, on a day-to-day basis." Looking at the anecdotal results, it's clear that objective was attained. Here's what managers reported back to HR:

- Employees are holding one another accountable
- People are taking ownership of their needs/engagement
- Employees are beginning to resolve issues and problem-solve on their own
- The energy conversations are saving managers time
- Sick time is dropping
- Overtime is dropping

"Traditionally, engagement has been seen as something that leaders need to own," says Ram. "But we saw employees beginning to own their own engagement – resolving their own issues and doing their own problem-solving. This saved managers time because they didn't have to be so involved in dealing with the day-to-day issues that employees could take on themselves."

Here's what's being reported to date from all engagement cohorts. The Energy Checks:

- "Allow me to take a deeper dive into what sometimes can be a difficult conversation"
- "Give me a greater sense of where my team really sits on the energy pendulum and then allow me to take action with the employees sooner to continue to keep them energized"
- "Give the team members more of an awareness of their own energy levels and when they might start feeling depleted"
- "Are helping us build higher levels of trust"
- Are self-fueling: "Staff come to me on their own asking for an Energy Check because they have something they need to discuss"
- Have become a day-to-day process: "This process is no longer a formal meeting but an informal part of everyday discussions"

The Juice engagement process has helped open people up more. It has started deeper discussions. We were told that:

- Employees are more active; they are helping each other out, not just doing their own work. The energy levels seem to be higher
- Employees are seeing that managers are more open and caring
- HR is seeing more self-awareness in staff and managers
- Managers are starting to see the benefits of revisiting areas of the engagement survey; this is not only opening discussions with staff now but is also setting up productive discussions when the current year's survey results are received
- In some cases, employees have seen that they are more energized than they perceived they were prior to reviewing the Energy Check questions and answering them

Throughout the process, Sandy Ram's goal has not wavered. "My vision is that, in the future, we're no longer holding an Energy Check tool in our hands to guide the conversation – that the energy conversation just becomes a natural part of our being and business operations."

As of this writing, Energy Check conversations have been rolled out across three significant parts of the organization – and, based on the results, Sandy has high hopes for sustainable success. Small wonder: they have achieved a 9 percent boost in their engagement scores, a significant achievement in anyone's books.

Energy Management Question

Should we be taking a survey sabbatical?

▶ ▶ ▶ Trust Conversations, Not Surveys

5

Seek Tension, Not Harmony

What's the brain science? Although the brain requires tension to do its best thinking, it perceives it as a threat to be avoided.

Where does this show up at work? Epic tensions emerge in the workplace, as can be illustrated by what emerges from employee engagement surveys. For example:

- Employees want better wages and benefits, but managers need to cut expenses
- Employees want more work-life balance, but managers need more discretionary effort from employees to drive business results
- Employees want to know under-performers are being held accountable, but managers are often muzzled and can't share the corrective actions they've taken

Because the natural response of the brain is to interpret tension as a threat, leaders become uncomfortable when these concerns emerge and will do everything in their power to remove the tension from the system.

Why does this matter? By attempting to remove tension, leaders miss out on the treasure that lies within tension: the most memorable, poignant, and energizing emotions that employees can have, which constitute a vast pool of free energy just waiting to be tapped. With training, leaders and managers can step in to employee tension and draw out the surprising innovations that turn competing priorities into sustainable solutions.

▶ ▶ ▶

Your Energy Pool

Imagine that you could position your organization over a large pool of renewable energy. Once hooked up, you're completely off the grid and power bills are a thing of the past. Think of the ways you could use that freed-up money to drive your business.

But imagine further that this energy pool could power up not only your physical facilities but your people as well, sparking innovations, fueling collaboration, animating meetings, and driving results.

You don't have to just imagine. There is such a pool of energy. It's called tension. From a physics standpoint, *wherever there is tension, energy is stored right inside it.* I can feel this energy when I walk through the departments of the best organizations. It's a kinetic buzz, an electric hum of productivity that fuels high performance.

And energy always seeks resolution. If the elastic in the image above could speak, it would say, "I'm being stretched thin – I want to be released!" Relational tension is similar. You don't get to decide whether energy will be released from the tension you have with another person. That's a given. But you do get to decide whether destructive energy or innovative energy is released. Your success as a leader is determined by how you release the energy inside tension.

All too often managers faced with tension-filled situations go binary and fulfill one of these drives at the expense of the other:

1 The manager who is all about pushing to get her needs met and doesn't consider the needs of her employees has slipped into "overpower" mode

2 The manager who is all about pulling out and meeting the employee's needs at the expense of getting her own needs met has slipped into "comply" mode

3 The manager who does not focus on her own or her employee's needs has slipped into "avoid" mode

These three reactions remove the one element required for breakthrough innovations: creative tension. So engagement solutions to surveys are simplistic, pendulum-swing, black-and-white fixes that create unintended consequences in other parts of the system.

Avoid, overpower, and comply reactions produce a crisis of belief in employees: "Will anything meaningful ever come out of this engagement thing?"

Cognitive Tension

Great teachers ask powerful questions. When they do, their students' brains clamor for a solution. Why? Because when you ask a great question it's like a bow is being pulled back inside the other person's mind. Tension is created. There's an arrow notched on that bowstring that's hungry for a bull's eye.

We'd never learn without tension. As Levy has written in his study of creativity, our brains are energized by it, interpreting it as a fascinating novelty – a puzzle to be resolved.

When your brain senses dissonance, it goes to work to resolve it. When it senses that something is missing, it goes to work to find it. Cognitive tension is brain fuel. In fact, if you never experienced tension, you'd never come up with a good idea.

As documented by Chan Kim and Renée Mauborgne in *Blue Ocean Strategy*, successful organizations like Southwest Airlines stepped in to tension and created uncontested market space that made the competition irrelevant.

The brains of leaders at Southwest were presented with cognitive tension: how to run an airline efficiently *and* delight your customers. Epic tensions always exist between the customer experience and the employee experience. In the airlines industry, passengers often expect unreasonable exceptions – for example, demanding that a gate agent severely bend the rules. Employees feel at risk. Should they uphold policy and face a threatening customer or concede to the customer and face a threatening boss?

At Southwest, wrestling with the creative tension between efficiency and customer service sparked an innovation that changed the game for employees and customers. According to Karan Girotra and Seerguei Netessine, on HBR Blog Network, the airline

> replaced hundred-page-long manuals of what the gate agent must do with a simple directive: "do whatever is necessary to get the plane out of the gate, fast." Certainly, from time to time employees will go overboard and spend too much time and money trying to get the job done, but it is easier to deal with these individual cases than to create constraining, enterprise-wide rules. The resulting productivity of employees at Southwest is several times higher than at any legacy airline, more than compensating for their higher salary.

We've been taught that necessity is the mother of invention. It's not. Tension is. The tension between the current way of doing things and the desired way of doing things is what sparks innovative thought. In short, the human brain requires tension to think creatively. As Georges Ghacibeh and his colleagues from the University of Florida neurology department point out: "Creativity has been defined as finding unity in what appears to be diversity."

Emotional Tension

But cognitive tension is only half of the equation when it comes to energizing the brain to do its best thinking. Emotional tension plays an equally vital role. In other words, it's not only the presence of two opposing thoughts that causes us to think innovatively, but two opposing emotions – specifically, the difference between what you want and what you have. As Robert Fritz has written, "This discrepancy forms a tension. Tension seeks resolution. The tension is a wonderful force because as it moves toward resolution, it generates energy that is useful in creating."

This is an important call-out. Tension is often associated with conflict, strife, and discord. But the emotional tension described above has nothing to do with strife. It is simply the powerful sense of motivation felt when we aspire to achieve something we're not yet experiencing.

To be clear, the tensions that release energy can stem from negative emotions like strife and angst, but they aren't restricted to those. In fact, the motivation that comes from positive forms of tension is what prods us into the hard slog of working through cognitive tension.

As we saw in chapter 1, the executive function of the brain is fuel-hungry and metabolically expensive. The brain needs a good reason to engage in difficult work. Emotional tension provides the necessary motivation.

The Treasures in Tension

Innovation isn't the only benefit of tension, and challenges regarding the customer experience aren't the only kind of tension.

When I ask groups to describe the finest moments of their history, they never fail to point to moments of epic tension.

"It was the SARS epidemic. We got out of our silos, came together, and achieved things we never thought possible."

"It was the flood in High River. Everybody pitched in. We were exhausted. We were super-energized."

Collaboration and identity are galvanized in times of tension.

Trust is another treasure hidden inside tension. Our brains code for trust in times of relational tension. Until we've gone through a tough issue with someone and discovered that they really have our back, our brain is unsure of whether we can trust them or not. When we work through the conflict and find they really have our best interest at heart, our brain codes them as trustworthy.

It's the tension between our desired future and our current reality that fuels our forward-moving efforts.

It's the tension between our perception of the challenge and our perception of our capability that we enter into a flow state – when the challenge is stretching us beyond what we believe we can do.

Another example of a treasure in tension is productive meetings. In *Death by Meeting* Patrick Lencioni says a good meeting should be like a good movie. The build-up of emotional tension and then the resolution of that tension release energy and make the experience memorable and poignant. That's one reason conflict is so vital in meetings. We need conflict and tension to create a sense of intrigue and energy. Without conflict, meetings become boring and lifeless.

Performance is another treasure found in tension. Teams can't truly perform until they learn how to storm.

Tension: The New Constant

Perhaps you're thinking, "I want more innovative energy. So where do I find some of this cognitive and emotional tension so I can tap in to it?" The answer is, *everywhere*. The three things every organization wants most are all in tension with one another: a great employee experience, a great customer experience, and great numbers.

- Become obsessed with creating the most memorable customer experience ever and you risk burning out your employees
- Become obsessed with protecting the work-life balance of your employees and you might fail to delight your customers
- Become obsessed with driving results and you could damage the customer and the employee experience

These three opposing drives represent the collective needs of everyone in your enterprise. Epic tensions will emerge over the things that people care about most: inclusion, information, power, results, and identity.

- Employees say, "We want to feel included in how things are done, but our suggestions seem to go into a black hole. Why do you even bother asking us?" Leaders, meanwhile, are commiserating with each other, saying, "These people just don't get the big picture. They don't understand that we don't have the time, money, energy, or capacity to execute on even half of this stuff"

- Some employees demand of their managers, "Share information with us," while others push back with, "Just tell us what we need to know – you're overwhelming our inboxes"
- Some leaders say, "Let's distribute control," while others insist, "No, we need to centralize it"
- The sales department says, "We need to create a differentiated customer experience," while Operations says, "We need global consistency to reduce costs"
- Branch offices plead, "Accommodate our unique cultural needs," and head office rebuts with, "No, we need to do what's right for the whole"

Every one of these examples represents a blend of cognitive and emotional tension. The bad news is that you will never run out of this kind of tension – it's the new constant in today's workplace. The good news is that it is a pool of renewable energy that will power up all the results that matter in your organization. You just need to learn how to release it.

The case story at the end of this chapter provides an example of a leader who stepped in to tension and pulled out astonishing energy – energy that transformed her entire organization. But, unfortunately, leaders often don't see the potential power within tension. In fact, the most common response leaders have to tension is to avoid it altogether. Let's look at both of these phenomena: avoidance and how to step in to tension.

Avoiding Tension

Jerome and Diane were senior leaders who had a big misunderstanding that just never got resolved. It was hurting business results because their blocked relationship created a bottleneck

– everybody had to go around them to get their needs met. Literally. Their offices were on the executive floor. Here's what the floor plan looked like.

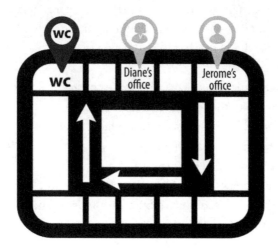

Each time Jerome needed to go to the washroom (four or five times a day), he faced a choice-point: walk past Diane's office or go the long way around. You can see the route he took. He walked the extra 100 meters because to walk past Diane's glass door was to risk the emotional tension of making eye contact with her.

One of the reasons employee engagement isn't working is that leaders follow a similar route of avoidance when presented with business tensions.

We worked with an organization that was paralyzed about how to respond to its survey results. Engagement scores in the sales force indicated that the reps were unhappy about compensation. Leaders were painfully aware that there was no money to sweeten the pot.

I asked them, "How are you going to respond to those concerns?"

"We're not," they told me. "We're not stepping in to that conversation with our sales reps. If we do, we'll create expectations we can never fulfill."

The tension was palpable, but the natural instinct of leaders was to avoid it. The internal narrative in situations like this is as predictable as it is illogical: "If we give this some time, maybe it will go away." You can see why employees are so cynical about engagement surveys. Why bother if there's no corporate will to step in to the tension?

Stepping in to Tension

Here's an example of how a large US-based organization turned the compensation conversation into a reward conversation and released energy out of an epic tension.

When the organization asked the sales force what their definition of feeling rewarded was, they discovered it was *freedom*: the latitude to shape their own schedules and set their own hours. "If I'm hitting my numbers, I want the freedom to knock off early and go golfing."

When freedom is a big driver for an employee, having the latitude to pick up their kids from school, take long weekends once in a while, or go golfing on Friday can mean more than the number of dollars on their paycheck. Reward is different from compensation. Stepping in to this conversation released the innovative energy out of the tension and gave leaders and sales people what they wanted: predictable results in a framework of flexibility. The leaders of the organization quickly ratified an agreement: "As long as you're meeting your numbers, you go golfing when you want to. It's all up to you."

What do you think happened? Sales reps worked hard for three or four days a week to hit their numbers, then used the other days to do whatever they wanted. Energy was released, the organization got its results, and the sales reps got their freedom.

Beyond Engagement is not just a management approach, it is a *way of being* that leaders demonstrate inside simple *energy conversations*. Our big work in the world is equipping managers like you to:

- Move toward tension
- Step in to the other person's world and draw out what matters most
- Help them see what matters most to you
- Identify the common ground – what you both want
- Harmonize the competing needs to release productive energy

When you use this approach to step in to the concerns that emerge from engagement surveys, you send a clear signal to employees that the engagement survey is doing exactly what it is intended to do: it is resolving tensions and releasing more energy. But you'll never be successful at that unless you get your two brains working in harmony.

The Tension Within: Your Two Remarkable Brains

You have an emotional brain and a rational brain. Each one has a vital, but different job to do. Your emotional brain's job is to keep you safe. Your rational brain's job is to serve up the best opportunity possible by analyzing all available information. Your emotional brain perceives tension as a threat. Your rational brain is fascinated by tension, seeing it as a novelty, a puzzle to be solved.

When these two brains work together in harmony, they give you exactly what you need to be a remarkable human being: a platform of safety and the processing power to solve life's tensions by holding two opposing thoughts side-by-side.

But there's a built-in challenge with these two brains: they function like a seesaw. So if your emotional brain believes you are in danger and becomes aroused, your rational brain – by virtue of the seesaw action – will become suppressed.

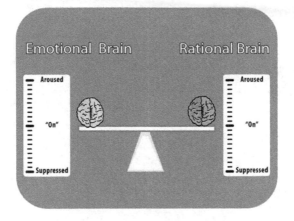

So tension exists not only all around you but also within you. As you'll see below, your emotional brain is intent on doing one thing, and your rational brain is intent on doing another. Great leaders learn to harmonize this internal tension and release the innovative energy out of it. As Daniel Siegel has put it, "There is evidence from current research that the most emotionally intelligent and effective people are those who can use the gifts of different parts of their brain together effectively; for example, empathy and intuition associated with the right hemisphere and logical and analytical skills in the left."

Let's take a look now at your two remarkable brains and what they can do for you.

As I said above, when these two brains work together in harmony, you can be your best self, stepping in to tension-filled situations and coming up with great solutions. But one of these brains is skittish about tension and reacts to it instantaneously and forcefully. You probably know from experience which one: your emotional brain.

You are alive to read these words today because your ancestors got really good at one skill: knowing when to fight and when to run. Whether they were confronted by a wooly mammoth, an enemy tribe, or a brushfire, they had to answer one question and get it right every time: Do we fight or take flight?

	Emotional Brain	Rational Brain
Other names	Old brain, limbic system, downstairs, reptilian brain	New brain, cognitive system, upstairs, human brain
Processing time	Reactive and lightning fast	Deliberate and slow(er)
Timeframe of operation	Always in the *now*	Able to engage in past and future as well as present thought
Information	Like a radar dish, it scans and receives information using cuedetection and social radar	Synthesizes, analyzes, and evaluates information
Interpretation	Tells you what is true based on past experiences – highly subjective	Scrutinizes assumptions with objectivity
Decision-making	Motivates the decision using gut instinct and intuition	Rationalizes the decision using logic and implication thinking
Type of reasoning	Social reasoning	Non-social reasoning
Level of connection with others	Highly attached through cuedetection, social radar, mirror neurons, and empathy	Detached
Focus	Safety	Opportunity

Your brain is no different. It is 100 percent committed to being absolutely right when it answers the fight-or-flight question. What conditions does it require to be right about that answer? Clarity, order, and simplicity. Your emotional brain loves the black-and-white. Why? Black-and-white thinking = speed.

Safety is job #1 for your emotional brain, and highly controlled conditions are what it needs to do its job. When things are clear, ordered, and simple, it has a platform from which to make the split-second decisions that keep you out of danger.

So you can see why tension spells threat to your emotional brain:

- "Too much complexity!"
- "Too much gray!"
- "Too many choices!"

- "Too much unknown!"
- "I'm losing control!"

These things cause rescue operations to kick in. In less than a second, a sequence of chemical events is triggered in your body that shuts down your slow, fuel-hungry rational brain and yields Central Command to your emotional brain. Immediate removal of threat is the primal need, so you proceed with lightning speed along increasingly rigid and extreme lines of thinking. In short, you go binary:

- "It's all or nothing"
- "It's now or never"
- "They're either good or evil – right or wrong"
- "You're either with me or against me"
- "You're either on the bus or off the bus!"

This type of thinking rarely pulls the treasure out of the tension. Why?

Because tension constricts the rational brain's processing capacity. In mere seconds, your ability to handle complex, future-based, analytical thought is reduced by up to 70 percent. In this condition, it is virtually impossible for you to hold two opposing views in your mind – which is precisely the work that's required to release innovative energy out of tension. As Arnsten has put it, "When the brain is influenced by the chemicals associated with stress, it is less creative and less able to think of long-term solutions."

How does tension create this energy drain? Through stress hormones that flood our system, shutting down future-based, complex thought.

As detailed by Rock and Page, research by professor and emotional intelligence expert Richard Boyatzis and colleagues found that when "research participants were being evaluated by others and therefore risked a reduction in status, their cortisol levels (an indicator of stress) remained higher for 50% longer."

High cortisol levels are simply a trouble light that signals what's going on inside us. "This tension is putting our security at risk. Back away from it!"

The following chapters explore how leaders and managers can stand right in the middle of tension without buckling to simplistic, binary thought. But let's mark this moment by saying that unless leaders and employees acquire the skills to self-regulate and manage their emotional brains, then like Jerome and the sales leaders above, they will go to any length to avoid tension. And when that happens, they'll miss out on the rich and rare opportunities that reside within it.

▶ CASE STORY
Tension Between Employee and Customer Experiences

Leslee Thompson stepped in to a maelstrom of tension when she was recruited as the new CEO of Kingston General Hospital. In 2008, KGH faced massive financial challenges, to the point that it was under provincial supervision. Public confidence in the hospital and staff morale were at an all-time low. The hospital's hand-hygiene compliance rates were some of the worst in the province, at just 34 percent. On top of all that, KGH consistently faced public scrutiny in the media in regular letters to the editor.

Six years later, things are different. KGH's hand-hygiene compliance stats are now some of the best in the province. Who helped hospital staff make that powerful shift? The very patients who, in some cases, had experienced harm at the hands of their caregivers. How did that happen?

Thompson picked up on a process of partnering with patients that began in the hospital in 2006, two years before she became CEO. She felt compelled to reach out to these patients and their family members. It was considered a bold move to invite parents and families to offer input at tables that had previously been "staff only," one that would create a power shift in the organization.

But Thompson and her team knew that to truly deal with the tension the hospital faced and transform KGH, she and her teams needed to turn the provider-centric focus on its head and start asking patients and families what their experience at KGH was really like – not what the hospital *thought* it was like, or *hoped* it was like, but what it *really* was like.

She engaged in a dialogue with 2,000 community members and listened to stories of how KGH had let people down, and what that felt like and how it had impacted their families. She acknowledged their stories and promised to work with them as partners to make things better in the future.

It wasn't a token promise. Within a short time, the Patient and Family Advisory Council, a group of former patients or family members of former patients, was formed. Besides improving the patient experience, the goal was to ensure that patients and their families' voices were included in every plan and decision made at the hospital.

Sure enough, their ideas and advice for dealing with KGH's challenges brought new ideas to the table, and difficult conversations to the forefront.

Leslee's instinct to step in to the tension led her to a massive pool of renewable energy. Those who had been through some of the most difficult hospital experiences said they wanted to make sure that what they or their loved ones experienced never happened to anyone else. "We *will* help you fix this." Patients and families who had positive experiences at KGH also wanted to help shape a better experience for other patients.

Thompson and her fellow leaders made a declaration: "For any decision in our hospital that has a material impact on the patient – a patient will be at the table."

These new champions received a role and a title: Patient Experience Advisors. As of this writing there are 60 of them. They have their own office in areas of prime real estate. They're in on the big decisions. A new wing doesn't get built and a new piece of equipment doesn't get purchased at KGH without their go-ahead. They also sit on hiring panels to ensure prospective candidates are patient- and family-centered. To date, KGH patients have helped pick more than 400 staff members, from nurses to senior managers to phlebotomists.

They currently sit on 77 active long-term and short-term committees at KGH. Through these committees they interview prospective employees; create and audit patient-care standards that will increase safety, quality, and patient satisfaction; design resources for cancer patients and caregivers; and brainstorm solutions to critical incident medical errors.

Some, like Marla Rosen, sit on multiple committees and volunteer for

40 hours a month. Last year alone, the 60 advisors volunteered 4,875 hours of their time, filling 165 advisory positions and helping 97 committees achieve their mandate. And best of all, their ongoing stream of innovations has made a material difference on the marker that matters most inside a hospital: patients are getting better, safer care.

Over the course of just six years, the hospital has undergone a remarkable financial, reputational and cultural transformation that KGH's advisors helped shape. It's a place the community has once again embraced as "our hospital," a place where staff's passion for health care has been reinvigorated, a place that is outperforming its peers and that is consistently named as a leader in patient- and family-centered care in national and international media.

The pool of free energy was always there for Kingston General Hospital; it just needed to be tapped.

Perhaps a pool of free energy is available to you. You are the one who gets to manage the energy that's stored inside tension. You get to influence the kind of energy that will be released.

And let me be clear. By "free energy" I do not mean getting employees to work for free – to come in early, pull all-nighters, and work weekends. No, I'm referring to energy that is bottled up in the normal days and normal capacities of normal people, energy that simply needs to be released. Your role as a manager is simple: release that energy.

Energy Management Question

Where have we done a good job of stepping in to tension and where have we avoided it?

▶ ▶ ▶ Seek Tension, Not Harmony

6

Practice Partnering, Not Parenting

What's the brain science? Our emotional brain perceives shared responsibility as a threat and triggers us to become under- or over-responsible.

Where does this show up at work? When time lines are tight, stakes are high, and the pressure is on to produce results, managers unwittingly slip from partnering into parenting.

Why does this matter? Parenting shuts down the two things that matter most: employees' willingness to offer their discretionary effort and their capacity to access their processing powers.

▶ ▶ ▶

From Parenting to Partnering

I have four kids. When they were infants, there was a lot of parenting going on and not much partnering. But now that they're adults, the opposite is true.

For example, when my son Adrian, now thirty-two, is renovating his home, I simply show up with my tools, and he tells me what project he's working on and how he plans to approach it. If

I have an idea that might add value, I offer it. Otherwise, I simply take his instructions and get to work.

There were myriad points over the past thirty-two years when my wife and I had to discern whether it was time to shift from parenting to partnering.

To get you thinking about the difference between these two approaches, let's start on the home front. From your point of view, which of the examples below is parenting and which is partnering? (You can do this whether you are a parent or an armchair parent.)

- I call my son's manager and tell him about how disappointed I am about the performance appraisal he's given my son
- I help my twenty-five-year-old son write a business plan for his new renovation business
- My seventeen-year-old daughter asks for my thoughts about the relationship issues she's facing, and I offer my perspective
- I get so disgusted with how filthy my ten-year-old daughter's room is that I clean it myself
- My twenty-four-year-old daughter (who has her own apartment) asks to stay over with Mom and Dad because she's lonely, and I say OK without asking any questions
- After reminding my son repeatedly that he needs to get the oil changed in his car, I finally take it to Jiffy Lube myself
- My son just got his first job and finishes at 1 a.m., and I offer to pick him up so he doesn't have to walk home or pay for a taxi

You probably laughed when you read the first sentence. But you would be shocked at the number of managers who say parents are asking to come along to their kid's job interviews or complaining about their performance appraisals. There's a lot of helicopter parenting going on: moms and dads who haven't been able to evolve from parent to partner.

Our job as parents is to help our kids move along the maturity continuum, to equip them to take increasing amounts of responsibility and ownership and expand their orbit of contribution in the world. This requires discerning which situations call for parenting and which for partnering.

There aren't many moms and dads who intentionally parent their adult children when partnering would be the superior choice. Nor are there many managers who think, as they eat their morning cornflakes, "How do I make my employees feel like children today?" But parenting happens all too often. What's the dynamic that triggers it?

The Binary Code of Responsibility

In my leadership sessions I ask leaders, "If you become over-responsible with finances at home, what does your spouse/partner do?"

They respond, "They become under-responsible."

"And if you become under-responsible in the area of taking care of the kids, what does your spouse do?"

"They become over-responsible."

What people know from experience, Roger Martin substantiates in his book *The Responsibility Virus*. As he puts it, "There is a relatively fixed amount of responsibility to be assumed in any situation, and any amount of over-responsibility is offset by an equal amount of under-responsibility."

In essence, you can view responsibility as a seesaw with an employee's under-responsibility triggering over-responsibility in the manager. The job of management is to partner with employees to bring them to a place of appropriate, ongoing, shared responsibility.

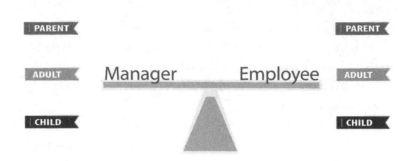

The school of psychology called Transactional Analysis brings this into an even sharper focus, showing that, as a manager. you can show up three ways in your relationship with an employee: as a parent, an adult, or a child.

Your employee can show up in the same three ways with you. It looks like this:

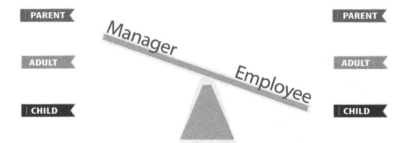

You want your dealings with employees to be adult to adult. But what happens when employees lose objectivity and feel completely stuck? They can slip into childlike behaviors, resisting change, creating cliques, complaining, gossiping, displaying a mindset of victimitis and entitlement.

How do you, as a manager, feel the impact of these under-responsible behaviors? Things don't get done, or they don't get done right. And how are you likely to react? By parenting your employees. After all, someone has to be responsible for the results in your department.

How does an enlightened, well-intentioned person like you turn into a parent? The answer is in your brain. As David Rock has shown, *your brain interprets shared responsibility as a form of threat and naturally avoids it.* You experience:

- Loss of control
- Lack of certainty
- Fear of failure
- Conflict, sparked by competing needs

Sharing responsibility with someone means relinquishing some of your ownership and control, and that can feel like a risk. Your brain has a simple solution: to go binary. Either you take all the responsibility, letting the other person off the hook but giving you ultimate control over the results, or you give all responsibility over to them, relinquishing control but getting off the hook in the risk department.

This binary approach to responsibility seems much safer to our brains, because sharing responsibility and partnering with someone means placing our reputation in the other person's hands, which makes us feel vulnerable.

Every person is wired with two healthy, but opposing drives: a push drive and a pull drive. The push drive enables us to be understood by others so we can get our needs met. The pull drive enables us to understand others so we can help them get their needs met.

The two drives are characterized by:

telling	asking
challenging	nurturing
declaring	inquiring
persuading	empathizing
being certain	being curious
Push Drive	**Pull Drive**

There are clearly times for each of these drives to operate in any healthy relationship, and they work remarkably well when they are blended. But just as very few of us are ambidextrous (one hand tends to be stronger and more coordinated than the other), very few of us bring an equal blend of push/pull to the situations in which we find ourselves.

For example, sometimes managers come across with too little pull and too much push, a parenting behavior I call "overpower." In other situations managers come across with too little push and too much pull, a parenting behavior I call "comply." Too little of either is an (absentee) parenting behavior I call "avoid." When you face tension-filled moments with employees, which one do you tend to slip into?

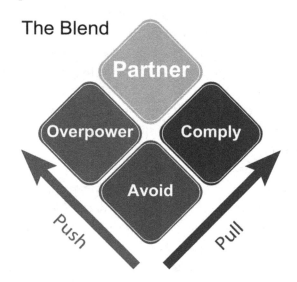

1 Managers who **overpower** (too much push and not enough pull):

   a Pressure their people to work without concern for the implications in their lives

   b Can come across as a drill-sergeant: "My way or the highway"

   c Demand unthinking obedience: "Because I said so"

   d Tell employees how they should do something before they even ask for help e Explain what they would do without asking employees what they've thought of so far

f Provide more information than necessary

g Over-remind employees

h Micro-manage and are slow to delegate

i Scold employees

j Pressure employees to "get on the bus" without giving them enough support

2 Managers who **comply** (too much pull and not enough push):

a Protect employees from the consequences of their actions

b Come across as helicoptering

c Rescue employees when they make or are about to make a mistake

d Comply with the concerns of employees, taking the monkey on their own back

e Give employees too much leeway in decisions, instead of weighing in

f Take work on their own shoulders: "It's easier just to do it myself. It needs to be done and I know they won't do it the right way"

g Are too concerned about the implications of work in employees' lives

h Enable employees and do not challenge them enough

3 Managers who **avoid** (not enough pull and not enough push):

a Give employees lots of space – in fact, they're never around

b Avoid issues until they become emergent

c Throw issues back over the fence to employees without engaging them

d Are not in touch with implications of work in their own world or their employees' world

e Fail to challenge and support employees

Regardless of what type of parenting behaviors managers slip into, they miss out on:

- Cultivating decision-making capability
- Calling people to higher levels of accountability
- Helping employees own and manage their own engagement
- Deepening relationships
- Unlocking results
- Releasing energy

Parenting: It's in Our Roots

Here's a two-minute history of Africa to illustrate how deeply parenting nestles inside humans' DNA.

Before 1850, Europe owned virtually none of Africa.

Then, around 1850, King Leopold II of Belgium made a move on the Congo for its rubber. The rest of Europe was already looking covetously at Africa for its abundant raw materials. When they observed Leopold taking what he wanted uncontested, they thought, "Cool. You can just go and take stuff and nobody stops you. Our guns trump their spears every time."

So the Berlin Conference was held in 1884 with European nations deciding how Africa would be carved up. By the way, some notable people were not invited. (Africans.)

By the late 1800s, virtually all of Africa was carved up and taken by the Europeans. The only exceptions were the countries of Liberia and Ethiopia.

The Europeans were expert colonizers. Everywhere they went they focused on the three Gs:

1 Goods

2 Glory

3 God

They took the *goods*: slaves, ivory, rubber, spices, tea, and diamonds.

They imposed their *glory*: their language, culture, customs, and education.

They brought their *god*: replacing the local religions with the religion of their homeland.

But colonizing wasn't a new idea. It didn't start in the mid-1800s with the Europeans. Colonialism characterizes the history of our world. As long as mankind has existed we have cast our eye on our neighbor's land and, if we were powerful enough or smart enough, have taken what we wanted and imposed our value system on the natives.

And we're still doing it today. Consider our attempts to lift "developing countries" out of poverty. The West has spent $2.3 trillion to better the fortunes of the third world. (Just for context, $2.3 trillion is a stack of thousand-dollar bills 248 kilometers high.) What has been accomplished with that stack of cash?

- Snowplows (yes, snowplows) are rusting at the Kampala airport
- Solar-powered ovens have become shelves and boxes, because nobody understood that most Africans cook at night
- The African landscape is checkered with dilapidated jeeps and water pumps that nobody was prepared to fix or get parts for
- AIDS-preventing condoms have been handed out like candy to men whose status is defined by their ability to sire children
- Malaria nets are being sold as wedding veils

Ironically, the countries that have received the most aid are the farthest behind, while those that have received the least aid are the farthest ahead. How can that be so? It's age-old colonialism tarted up as foreign aid. "We know what's best for you, and we're here to save you."

What's the essence of the colonizing mindset? It's parenting, not partnering – and it can best be understood by unpacking the roots of parenting by looking at the related words *paternalism* and *patronizing*.

The essence of paternalism is acting "as if you were my child." The essence of patronizing is acting "as if you were my servant." In both cases the essential meaning is the same: *you're missing something, and I need to supply it*. Canadians should be familiar with this: it's how we have treated First Nations people.

You're missing something and I need to supply it is the essence of parenting and has been bred into every one of us over thousands of years of colonization. It always produces the same result: a learned helplessness based on the messages we receive:

- You're missing the intelligence so I need to supply my advice
- You're missing the motivation so I need to supply you with my reminders
- You're missing the capability so it's quicker for me to do it myself

Moving beyond engagement is pretty much impossible with these kinds of approaches. They show up most clearly when managers try to address their teams' engagement scores. Sitting with other leaders or HR partners, they come up with fixes that put them squarely in the parenting role – which decreases engagement as employees are kept from managing their own energy and owning their own engagement.

We're not talking about bad managers/bad leaders here but about what happens when our brain's fight/flight wiring combines with our colonizing DNA. We take identity in being *helpful*. We take identity in being *nice, hard working, productive*. We get a buzz out of solving people's problems.

Thinking back to chapter 3, parenting produces a predictable outcome: intended support, not felt support. Parented in ways that feel inappropriate to the circumstance, employees experience the three rogue elements that shut down intelligence and discretionary effort:

1 Not enough safety
2 Not enough challenge
3 Not enough care

The impact is predictable: not enough innovative energy.

But what if employees already have what we think is missing? What if they already have the desire to make a difference, the intelligence to generate great ideas, and pride in doing a great job? What if they come with their own forms of self-determination already baked in: a good work ethic, values, character? Imagine if your job as a manager was not to get people to go the extra mile but to build the conditions in which there was nothing standing in the way of their doing so.

Partnering Is the Goal

Am I saying that parenting is always a bad thing? Is there never a time to parent at work? Clearly there is. Affirming, supporting, directing, encouraging, advising, and praising could all be described as parenting behaviors, yet each of them plays a vital role in the workplace. So what am I saying with this call-out to practice partnering, not parenting?

Some psychologists say that many of us need to be re-parented. Few of us had perfect parents, and parts of us were not parented as well as they could have been. For example, some employees have a childlike need for affirmation, others for inclusion, and others for control.

At the beginning of their journey with you, it will be appropriate for you to offer them a bit more attention, affirmation, and protection than you will at the end. When I ask people to tell me about the best leader they've ever worked for, they frequently tell me stories of their first boss. "I was so green. He took me under his wing and showed me the ropes. He really had my back. He was like a father to me."

There is a time for parenting. When? Well, management is seldom black and white. Life is circumstantial. We work in the grays. But you can learn to identify the felt needs of your employee and work with them to get those needs met in a way that provides parenting but clearly feels like partnering.

So while parenting can be necessary and appropriate, partnering is the goal. The key is to be aware when you are slipping into the default of parenting when partnering is what's being called for.

The Partnering Diagnostic

While you as a manager can be trained and coached in the skills of awareness and self-regulation, how do you retain these skills in tension-filled moments with an employee? *By using the partnering diagnostic of holding out for each other's highest good.*

Highest Good

It takes two to get to the highest good in a relationship. If you and I are partnering, I don't know what your highest good is apart from you. But you don't know what your highest good is without me, because you can't see yourself objectively. You've developed self-limiting beliefs.

Face it, you're the project you've been working on your whole life, and that can get discouraging. It's easy to lose hope and optimism about your highest good. But I see you more objectively. I see glimpses of your highest good and your true potential, even if you've lost sight of them. In short, the highest good for both of us emerges only through a partnering dialogue.

Holding Out for the Highest Good

But it takes more than just seeing the highest good. We have to hold out for it, cultivate it, protect it, fight for it. And it's always worth holding out for, because a person's highest good is a combination of the two things that matter most: the relationships they build and the results they generate. So partnering is two people contracting to hold each other accountable for their impact on results and their impact on relationships.

Let's say I'm the relationship guy, living in harmony with all and loved by everybody but not hitting my numbers. You hold out for my highest good by holding me accountable for my impact on results.

On the other hand, let's say you're the results maven, technically brilliant and getting things done but trampling people in the process. I hold out for your highest good by holding you accountable for your impact on relationships.

Your ability to partner with employees to solve engagement issues is what lifts your burden as a leader or manager, freeing you from your unsustainable parenting roles. It is well worth making the shift from parenting to partnering.

To get down to specifics, here's an illustration of how managers can resolve tricky tensions by holding out for their workers' highest good.

▶ CASE STORY
A New Time-Off Policy

In chapter 2 I discussed how we helped Don Ford and his senior team boost engagement scores at the health-care access center they lead by focusing on the employee experience. Here, I'd like to show you a different dimension of their journey: the shift they made from parenting to partnering.

Stephen Kay, the VP of human resources, wanted to discover the answer to one question: "If there was one thing you could change today, what would it be?" Juice Inc. ran focus groups across the organization to draw out and identify the answer to that question.

In group after group, the answer came out with singular clarity and passionate force: "Change the time-off policy." Employees felt that the policy was causing them to be treated like irresponsible children. It was destroying trust and making them feel controlled and micro-managed.

They frequently had to put in extra, unpaid hours on behalf of the organization, and they did so willingly. Yet there was no give and take. If they needed an hour off to go to an appointment, it was an exercise in pulling teeth to get permission. For instance, if Janet needed time to pick up her aunt who was flying in from England, she basically had to take a sick day to make it happen.

I went to the senior team with the results of the focus groups and shared that several themes had emerged but the burning issue that needed to be addressed was this time-off policy.

"Ah, yes," they said, "that's one of our big business tensions. The employees want flexibility, but we need consistency. We'd love to give them time off whenever they need it, but if we do, it could damage the service we give our clients."

But the senior leaders were committed to respond to the output from the focus groups, so they said, "We will fix this."

My response raised some eyebrows.

"Please do not fix this," I said. "That will fly in the face of everything we're trying to do here."

I asked them to consider an important question – the question I mentioned above that enables you to know whether you are parenting or partnering in any given situation: "What will lead to the highest good for this person?"

When we worked through that question, it became clear that asking employees to help fix this issue would promote the highest good.

An invitation went out to employees, and the uptake was strong. A working committee was formed made up of nineteen employees and three managers. They worked long and hard to draw out what mattered most to managers and what mattered most to employees and come up with a solution they thought would work.

And it did work. Six months after the new policy was introduced, an anonymous survey went out to the organization to see what people thought of it. A whopping 98 percent of respondents said, "Don't change it. It's a great policy."

Are you curious about the policy they came up with that elicited that kind of response? It centered on two brilliant concepts:

- Promoting the accountability of employees to use freedom responsibly
- Unburdening managers of the time-eating, soul-sucking decisions about "fairness"

The working committee asked themselves an important question: "If Janet needs Friday afternoon off to pick up her aunt, who will be impacted?"

The answer was simple: her clients and her teammates. So the new policy was grounded in respect for those two stakeholders. Here's how it went.

If Janet needs time off to pick up her aunt, she first makes arrangements to ensure her files and cases are well ordered, so anybody else could step in and take care of things in her absence. Check.

Then, at a team meeting, she asks the team if they can cover for her. Team members look at their schedules, talk about it, and say, "Yes." Double-check.

Because health care is still bureaucratic, the manager still has to give the final OK. But when the employee already has the thumbs up from her team, how much mind-space does the manager need to give to this decision? Zero. It simply has to be rubber-stamped. Triple-check.

You can imagine the amount of time this has freed up for managers. No longer do they have to be time-off referees. You know the drill: "I'd love to do this for you, but if I do it will set a precedent and then everybody will…" Instead, they do what they're supposed to do: run the business.

You can also imagine the toxic acrimony and resentment that was lanced with this elegant move.

And employees have become more accountable. They know that if they abuse the system, it will be hard to get coverage the next time they need it. And, best of all, they own the policy. Why wouldn't they? It's theirs. This is a great example of free energy. The great ideas and hard efforts were bottled up inside the employees.

They simply needed the invitation, "Partner with us."

Energy Management Question

In what ways are we *parenting* and in what ways are we *partnering* with our employees?

▶ ▶ ▶ *Practice Partnering, Not Parenting*

7

Pull Out the Backstory, Not the Action Plan

What's the brain science? Conversation is the operating system (O/S) of our brain.

Brain activity, at its most elemental level, consists of neuron A transmitting a message to neuron B and neuron B receiving the message. This simple neuronal conversation is the basic unit of communication, allowing our software (our brains) to talk to our hardware (our bodies).

Remove the conversation between individual neurons, and the human body would cease to function. Individual neurons do not feel or think. Their power is through their connection to other neurons.

But conversation is not only the O/S within our brains; it is also the O/S *between* our brains. Researchers are showing that these neuronal conversations within our brains are significantly strengthened and enhanced by other people's neuronal conversations – by social conversations "between brains."

Where does this show up at work? Leaders who fail to understand this forego conversations with employees in favor of relying on the employee engagement survey data. By the time they hash out their engagement strategies and present them to employees, they're serving up "reality" that is three, six, or even nine months old. Stale intel makes employees turn up their noses and tune out.

But the issue is bigger than the real-time integrity of your intelligence. You're trusting in engagement strategies to boost engagement, and they seldom do. There's no blanket solution for engagement – it happens one conversation at a time.

Conversation – sitting with employees and pulling out the backstory, the deeper context contributing to the survey results – is a vital step, not just because it identifies the most intelligent solutions but because it syncs with the way your employees' brains work.

Implementing a solution without having a conversation with employees short-circuits their inner wiring and practically guarantees non-adherence or even covert resistance.

Why does this matter? Any engagement activity that is not conversation-driven runs the risk of coming across as parenting, not partnering. Leaders make engagement work when they use simple conversations to get to the backstory and draw out what matters most to employees. It produces a powerful impact as employees begin to manage their own engagement.

▶ ▶ ▶

Your Conversational O/S

Researchers such as Nasuto and de Meyer, as well as Daniel Amit, have explained that the operating system of the brain consists of the basic unit of communication – neuron A transmitting information to neuron B, and neuron B receiving it – multiplied a multitude of times across a lifetime. Your brain is a network of 100 billion neurons, and each one has conversations with up to a thousand of its neighbors. When you think of it, there are more conversations going on inside your head on any given day than happen verbally in the entire world. That's quite a neuronal community inside your cranium.

An operating system has a simple purpose: as stated by howstuffworks.com, it "organizes and controls hardware and software so the device it lives in can behave in a predictable but flexible way." Your brain uses neuronal conversations as an O/S to power up speech, sight, and movement – all the sexy apps that enhance your life.

I'm not saying that conversation is Central Command – the control center of the brain. Neuroscientists are unable to agree about which part of the brain is CEO. Some say it's the neo-cortex, others say it's the reticular activating system (RAS). Still others say it's the "higher I," or spiritual self.

Of course the role of neuronal conversation goes farther and deeper than our ability to see, speak, and move. Our conversational O/S drives a more elemental app: our thinking. In other words, thought is a conversation-driven process.

Just a few minutes ago I was calculating the amount of board and batten I needed to clad my cottage. I was taking measurements and voicing my calculations out loud. "OK, 35 feet long times an average of 8 feet high is…8 times 30 is 240 plus 8 times 5 is 40. So that's 280 square feet of board."

Perhaps you've caught yourself doing something similar. If so, don't worry. You may be normal. Research shows that *mathematical and verbal problem-solving are accompanied by covert oral movement of the tongue and lips.*

In other words, when we need to problem-solve, we instinctively begin a conversation, always internally (inside our head) and sometimes even externally (our tongue and lips get involved).

That people don't process well without internal or external conversation is a big problem when it comes to current employee engagement practices. Why? Because they are relatively conversation-less. Which is a big reason why engagement isn't working. These practices actually stymy the energy that comes from people connecting in ways that increase their energy and their outcomes. Knowing this, you have a great opportunity to leverage conversation to sync with the way your employees think – which is... socially.

There Is No Brain but the Social Brain

Matthew Lieberman, one of the world's foremost authorities on the study of neuroscience, has uncovered a fact with far-reaching implications for those of us who care about employee engagement. The first and foremost priority of the human brain, he says, is social cognition, "making sense of other people and ourselves."

If your job is people, I recommend you read his book *Social*. If you do, you'll see how our need for connection is greater than our need for food and shelter. You'll discover that the brain's favorite activity is mind-reading: predicting and anticipating the reactions of others. And you'll learn that the brain's preoccupation with connection and mind-reading enable it do the one activity that has allowed us to survive and thrive as a species: cooperate with others.

Our brains, Lieberman says, "were wired for...*reaching out to and interacting with others.*"

Which is precisely what conversation is all about. It's our native wiring, the perfect operating system we need to connect, understand, and harmonize with others.

It's no wonder, then, that more than 95 percent of people report having friends. And it's no wonder that over a billion people have Facebook accounts and that more people visit Facebook than any website in the world (including Google). Facebook gives us a way of doing the three things we crave most: connecting with people, getting a chance to read their minds, and coordinating activities with them.

Sociality: How Our Brains Developed

Not only did social cognition help us survive as a species and not only does it motivate most of our current behavior, but being social actually developed our brains in the first place. As D. Franks puts it in *Neurosociology*:

> While our individual brains are singular and self-contained, the processes on which they depend for functioning are social ones. We have seen that there is no fully working human brain without the presence of other brains. The functioning brain is social in the sense that any given brain is completely dependent on other brains for its development. Without question, the synaptic brain is contained in our individual skulls but the intangible thought processes which these synapses make possible depend on a social environment with other actors who are engaged in everyday public discourse and interaction.

Even at this very moment your brain is being influenced and infused with cultural norms, popular ideas, and societal trends. As Leandro Herrero writes in *Homo Imitans*, imitation isn't just the way we learn to walk and talk; it's the way we operate. Think about the shifts we've seen regarding seatbelts, littering, drinking and driving, and the environment. Every societal transformation relies on imitation for its success.

As S. W. Gregory, Jr., has put it, at both the conscious and the unconscious level, "we are social to the core."

Our Organizations Have an O/S

As conversation is the O/S of our brains, so it is the O/S of our organizations. And as the neuron is the basic unit of the brain, so the employee is the basic unit of the organization. As neuronal conversations are the way the brain gets things done, so employee conversations are the way the organization gets things done.

Conversation is the O/S that enables the apps in an organization to work: customer service, sales, feedback, coaching, strategy, innovation. Fantastic apps, but what happens, for example, to customer service if the operating system of conversation crashes? It's disabled.

Conversation is what binds all the hardware (facilities, technology, etc.) and software (people, IP, etc.) in an organization and allows them to function flexibly and predictably.

Moving beyond engagement is about honoring our native wiring, which needs to utilize conversation to drive the apps that make things easier and produce great results. Short, simple energy conversations power up sales, customer service, innovation, coaching, and feedback.

And when it comes to responding to engagement survey results, conversation is the O/S that enables you to pull out employees' backstory, the key to releasing their energy.

What's a Backstory?

If you were cast for a role in a major film, you'd be given what's called the backstory to help you get inside the skin of your character. The backstory gives you context: important details about events, background information about relationships, and subtle nuances about the emotional needs and values that matter most to your character.

It brings your character to life. Without the backstory, your portrayal would be one-dimensional at best.

And as you'll see, without the backstory, your response to your employees' engagement scores will come across as wooden and un-magical.

"Buy Picnic Tables!"

We trained all the store managers of one of Canada's largest grocery retailers and taught them how to step in to their employees' worlds to find out what matters most. Our training was specifically designed to correct a major flaw in the managers' approach:

looking at engagement scores and assuming they knew how to respond to them.

One store had survey scores that indicated the issue of respect was of crucial concern to employees. We happened to know the backstory behind the low respect scores. Employees really only wanted one thing from their manager: they wanted him, when he did his walk-through at the store first thing in the morning, to acknowledge them with a greeting and call them by name, even if he had to read their name off their name badge.

But that wasn't happening. Instead, the manager would walk in to the produce section and, without greeting the employee, would bark out, "What's that wet floor sign doing on the floor? It's not raining any more!" or "How come there's a spelling mistake in that signage?"

And, turning to leave, he'd add, "And what's that jar of salsa doing in with the cucumbers?"

This clearly left something to be desired when it came to the employee experience. Instead of pulling out the backstory, the manager looked at the low respect scores, assumed he knew how to solve them, and said to his assistant manager, "I know exactly what we need to do: buy picnic tables! That way people will have a place to go to have a smoke or eat their lunch outside. This will make them feel respected."

As you can imagine, the picnic tables were the laughing stock of the store for months to come.

Has your organization presented any "picnic tables" in response to engagement surveys? If so, the following story can help you learn something about drawing out the backstory that could make a big difference to your engagement scores.

Conversation syncs with how people think. Because of that, it's an operating system that costs little and yields much. But you need to be intentional as a leader to integrate that operating system into your engagement initiatives to uncover the backstory of employees. What would this look like in your organization? Here's how it looked at Yum! Brands.

► **CASE STORY**

Helping Employees Own Their Own Engagement

Fortune 500 business Yum! Brands, Inc. is the world's leading restaurant company with brands like KFC, Pizza Hut, and Taco Bell. With 1.5 million associates worldwide, Yum! operates over 40,000 restaurants across more than 125 countries and territories.

Yum! leaders put a strong emphasis on people: ensuring that employees know they can make a difference and frequently infusing the corporate culture with energy, opportunity, and fun.

And the organization's commitment to employees has not gone unrecognized: in 2013, Yum! Restaurants International (Canada) was ranked as one of the nation's best places to work by Great Place to Work Institute Canada.

The results were extremely positive, yet the company's Canadian leadership wanted to do better.

"With an 85 percent average employee engagement level, we knew we had a good culture at our Yum! Canada offices – but we didn't want to let our guards down based on those results alone," says Brian Henry, the company's Senior Director in Canada.

"It simply wasn't enough for us to work on all the 'low-hanging fruit' tactical things: we wanted to see year-over-year results," says Brian. "If, year-over-year, we could say that we'd seen more than a 5 percent improvement in any category, we would know we were creating the right level of energy with employees – and responding to their needs."

Human Resources traditionally owned and was accountable for employee engagement. Yum!'s leadership believed some employees were holding back on their opinions pertaining to the corporate culture.

"Our HR team is very close to employees, but generally speaking, it is often seen as distanced and unaware of how employees really feel," says Brian. "We began to realize that employees were the best people to say what was – and wasn't – going right."

As a result, in early 2013 Brian and his team began to explore how to minimize HR's role in corporate action planning. "In other words: How could we flip things upside-down and have Yum! employees become more accountable for their own engagement?" – a perfect example of shifting from parenting to partnering.

The PowerUp Squad

To best address issues identified in the Great Place to Work survey, Yum! created an action-planning team of six employees. Named the PowerUp Squad, this advocate group was developed to be a cross-functional team representing each Yum! brand, function, and job level – and to remove any barriers existing between HR and the overall organization.

"Each team member is embedded within teams to the point where they truly understand the day-to-day life of employees," says Marzena Dodolak, Senior Manager of HR for the company in Canada. "Because of their positions in the company, they could penetrate far deeper within the 'employee trenches' and build credibility more quickly among their peers than an HR-centric group."

By early 2013, the PowerUp Squad was fully set up and ready to take action as a conduit for ideas and solutions from all levels of the company. But the question remained: how, and where, would they start?

Squad Training

Juice helped the PowerUp Squad develop the right mindset – particularly how to most effectively tackle such a large project as earning employee trust through simple, straightforward conversations. Participants learned the skills required to effectively step in to employees' worlds to:

- Draw out what matters most
- Manage any tensions that exist
- Harmonize competing ideas and priorities

Our training offered a springboard so the PowerUp Squad could figure out the best ways to engage with employees. Says Marzena, "Squad members learned how to create and manage focus groups, framing questions to get employees to tell them what they really thought about particular workplace issues."

"At the end of training, we saw two major benefits," says Brian.

"First, it allowed our squad to be more effective as a team. Second, we saw Juice's training as an incredibly helpful development tool to help our PowerUp Squad members take on leadership roles in anything transformation-related."

The Results

A year since the it was created, employees continue to approach the Yum! PowerUp Squad.

"Ideas are coming from everywhere now," says Brian. "Of course, we are still receiving feedback directly through Human Resources, but creating the PowerUp Squad has resulted in a great increase of employee opinions and engagement."

"Meanwhile, HR is being influenced by the learnings of the PowerUp Squad," adds Marzena. "What we now have is an extremely fluid relationship between Human Resources and the squad."

As well, Yum!'s efforts to improve year-over-year results have paid off. In 2014, the company saw employee trust rise by 7 percent on average. Moreover, Yum! scores went up on nearly 90 percent of the statements of the Great Place to Work survey.

Energy Management Question

What would change for us if we began to treat conversation as our operating system?

▶ ▶ ▶ Pull Out the Backstory, Not the Action Plan

8

Think Sticks, Not Carrots

What's the brain science? Carrots – the offer of future pleasure or of the meeting of needs – have nothing on sticks when it comes to commanding the brain's attention, which, through the amygdala, is on high alert 24/7 for the tiniest cues of danger.

There are few drives more important to the success of your organization than the universal human drive for self-efficacy, the belief that "I am capable of doing what it takes to achieve results." Recent research emerging from *The Progress Principle* by Amabile and Kramer demonstrates that, "of all the positive events that influence inner work life, the single most powerful is progress in meaningful work."

Where does this show up at work? Leaders often gravitate to offering carrots – recognition, cheerleading, and inspiration – instead of thinking sticks, that is, looking for and addressing the psychological forms of interference that undo employees' best efforts.

When employees experience blockages, barriers, and setbacks that short-circuit their ability to achieve goals, self-doubt sets in and they begin to second-guess themselves.

Why does this matter? Negative events at work have far more impact on people's performance than positive events. Failing to realize this, leaders gravitate to cheerleading activities like recognition programs, incentive plans, inspirational town halls, and team-building events to boost their engagement scores.

▶ ▶ ▶

Highly Engaged/Highly Frustrated

Take a look at the equation below. What happens when you take a highly committed superstar who is deeply wired to make a difference and put her in a situation where under-performers aren't dealt with, systems aren't working properly, and bureaucracy rules? You guessed it: frustration. In a culture where interference is not addressed, the highly engaged become the highly frustrated.

Research conducted by the Hay Group (using a database of four million) reports that between 32 and 48 percent of employees worldwide report work conditions that do not allow them to be as productive as they could be.

How far-reaching is the frustration equation? Further studies by the Hay Group show that 20 percent of today's workers are frustrated. What's the percentage in your organization?

Motivational carrot tactics get employees to try harder for a short while, but if interference is not removed, employees come to see these tactics as pay-off bids.

Big gains occur when managers learn how to move toward tension and address the sticks, partnering with employees to remove the interference that short-circuits performance and depletes energy. Let's explore why sticks are so important to employees.

The Gift of Fear

"I walked into that convenience store to buy a few magazines and for some reason, I was suddenly…afraid, and I turned right around and walked out. I don't know what told me to leave, but later that day I heard about the shooting."

The relieved survivor is telling his story to a pre-eminent expert on violence and how people can steer clear of it. The expert's name is Gavin de Becker, a three-time U.S. presidential appointee whose pioneering work changed the way governments evaluate threats to their highest officials. In his book *The Gift of Fear* de Becker relates how he asked the man what he saw – what he reacted to. The response:

> Nothing, it was just a gut feeling. [A pause.] Well, now that I think back, the guy behind the counter looked at me with a very rapid glance, just jerked his head toward me for an instant, and I guess I'm used to the clerk sizing you up when you walk in, but he was intently looking at another customer, and that must have seemed odd to me. I must have seen that he was concerned.

De Becker teaches people that what they easily dismiss as gut feeling or intuition, in fact is "a cognitive process, faster than we recognize and far different from the familiar step-by-step thinking we rely on so willingly." He notes that most people "think conscious thought is somehow better, when in fact, intuition is soaring flight compared to the plodding of logic."

"Nature's greatest accomplishment, the human brain," he continues, "is never more efficient or invested than when its host is at risk. Then, intuition is catapulted to another level entirely, a height at which it can accurately be called graceful, even miraculous.

Intuition is the journey from A to Z without stopping at any other letter along the way. It is knowing without knowing why."

There it is. Risk catapults intuition to miraculous levels. In short, your brain gives the nod to carrots, but is hyper-vigilant about sticks. Why? **Because stick-centricity keeps you alive**.

Why We Like to Be Right

You are reading this paragraph today because your ancestors were able to answer one question and get it absolutely right, every single time. The question? "Do we run or do we fight?"

A fur-clad scout ran up to your great, great, great, etc. grandfather and stammered, "The neighboring tribe is here to take all our women away. Do we run or fight?"

Your great, great, great, etc. grandfather quickly compared the fighting prowess of his opponent with that of his tribe, and pronounced, "We fight!" If he hadn't gotten that answer absolutely right, your great, great, great, etc. grandmother would have been sported away and you would not be here today.

There's a part of your brain, the amygdala, that is like a massive radar dish, hyper-vigilantly seeking out the slightest sign of threat. The amygdala is paranoid – on high alert 24/7 – even as you sleep, detecting the tiniest cues of danger.

But you're not just triggered by physical dangers like bears, snakes, and criminal assailants. Psychological threats send you into high alert, too. Any time you feel like your significance, belonging, or freedom are at risk, you jump to rescue operations and lose a large percentage of your executive function. Your brain pays strict attention to psychological sticks. And that makes all the difference in the world when it comes to employee engagement and performance.

What Blocks Performance?

When managers witness a dip in an employee's performance, they often ask themselves the wrong question: "What's missing here

– knowledge, effort, talent, commitment?" But as it turns out, if they have made a good hire, a lack of things is seldom the problem. Most engagement processes focus on adding something for an incremental gain. But sustainable engagement is actually about removing something for an exponential gain. Let's see what that something is.

Let's say you have an employee named Lindsay. You hired her because she was the perfect package of talent, knowledge, experience, education, skills, and character. The conventional wisdom around performance (P) is an equation that looks like this:

This equation would lead you to believe that with all these traits, getting performance out of Lindsay will pretty much be a slam-dunk. But there is a missing variable in this equation – as it turns out, a very vital one: Lindsay divided by interference (I).

When you think back on your life, there probably were times when you were all you could be and other times when you were less than you could be. At any moment of the day, every one of your employees is performing in one of these states. Think of the states in terms of your own capacity.

Full Capacity – This is what you can be and do when nothing is in your way, when you're tapping all your latent, inherent capabilities. You're equal to the task, functioning at your full ability to perform. In this state you can access all that's yours: your knowledge, energy, experience, skills, and strengths.

Below Capacity – This is you operating at less than your actual ability to perform. You can't do what you normally can do, or be what you normally can be. In this state, you're not able to access all of your knowledge, energy, experience, skills, or strengths.

As you read on, you will see:

1 The biggest cause for people working *below* capacity.
2 What you can do to help people access all that's theirs – to work at *full* capacity.

How Below-Capacity Happens

We were working with a Chief Nursing Officer (CNO) who told us, "Nine out of ten of our surgeons are fabulous to work with, but one out of ten can create huge havoc for us. We've got an ER surgeon who is technically brilliant, but he's impatient, demeaning, barks out orders, and explodes at nurses.

"I've got nurses who are talented, energized, and knowledgeable, but when they work with that surgeon they are so on edge they make dumb mistakes."

The CNO told us their performance was dipping not because of a lack of talent, energy, or knowledge, but because of something else that I identified above as the missing variable, interference.

In this instance, the interference is self-doubt created by the fear of being attacked in front of one's peers. Interference is the great fractionalizer. It's so powerful it can fractionalize (read divide) Lindsay's talents, short-circuit her knowledge, and make her experience and skills irrelevant. She is much less than she could be.

Why is interference so powerful at dividing performance? Because your amygdala is highly reactive to any sense of threat. When triggered, it shuts down 70 percent of your rational processing powers, sending you into binary, black-and-white knee-jerk reactions and rescue operations. That response is good for survival but bad for deploying your talents, knowledge, and skills.

Imagine the impact of interference in this team meeting (from Pearson and Porath's book *The Cost of Bad Behavior*):

> We were having our daily planning meeting, and things weren't going so well. We were off on our projected numbers, and everybody knew it. At the beginning of the meeting the manager passed out job applications for Wal-Mart. He explained that if we didn't start making our numbers immediately, several of us were going to be looking for new jobs soon.

How focused do you think these employees were after they suffered that psychological poke in the eye?

The forms of interference that fractionalize performance are myriad, but the biggest ones are psychological, like these:

- Working in a state of fear, due to the threat of being bullied
- Self-doubt stemming from feeling judged or evaluated
- Unresolved conflict, team tension
- Being ostracized or excluded
- Feeling unappreciated or unduly criticized
- Carrying more than one's share because under-performers are not being dealt with
- Being in "overwhelm mode," juggling too many competing priorities
- Lack of clarity around what is expected

This list represents only a tiny sliver of the many forms of interference that exist. It is fairly safe to bet that every single employee, manager, and leader in your organization encounters at least one kind of interference every day on the job (and at home as well).

The key point, as researched by Baumeister et al., is this: Negative events like the ones above are far more powerful than positive events when it comes to shaping the lived experience of employees. Consider these four examples quoted from *The Progress Principle* by Amabile and Kramer:

- The effect of setbacks on emotions is stronger than the effect of progress.
- Although progress increases happiness and decreases frustration, the effect of setbacks is not only opposite on both types of emotions, it is greater. The power of setbacks to diminish happiness is more than twice as strong as the power of progress to boost happiness. The power of setbacks to increase frustration is more than three times as strong as the power of progress to decrease frustration.
- The connection between mood and negative work events is about five times stronger than the connection between mood and positive events.
- Employees recall more negative leader actions than positive ones, and they recall the negative actions more intensely and in more detail than the positive ones.

As Mizuno et al. have written, the amount of human energy depleted by interference is difficult to calculate, but one thing is certain: energy is depleted. Science explains why. Basile et al. have shown that, when your brain experiences interference, your sympathetic nervous system (SNS) is triggered and your energy stores are rapidly depleted. But when your brain senses interference being removed, its energy conservation system (PNS) kicks in and your energy stores are rejuvenated.

The Cost of Interference

The incivilities caused by unaware or toxic managers produce a wake of interference that has a significant cost. A study of 775 Cisco Systems employees who had experienced workplace incivility revealed (and I quote in full):

- Fifty-three percent of employees surveyed lost work time worrying about the incident and future interactions with the offender.
- Twenty-eight percent lost work time trying to avoid the offender.
- Thirty-seven percent reported a weakened sense of commitment to their organizations
- Twenty-two percent reduced their efforts at work.
- Ten percent decreased the amount of time they spent at work.
- Forty-six percent thought about changing their jobs – to get away from the offender.
- Twelve percent actually changed jobs.

These data reveal a chilling truth. Your best employees can be the ones whose performance is most disrupted by interference, and they often suffer this experience at the hands of the oblivious, the unskillful, and, sometimes, the malicious. You'll never make engagement work unless you learn how to reduce or remove this energy-sucking virus.

What Makes Interference, Interference?

Think back to the operating room nurses working with the explosive surgeon. The same behavior that creates debilitating interference for Lindsay rolls right off the back of her co-worker Danelka. Why? Because it's their driving *need* plus what they *believe* that makes something interference or not. Here's a short example of how the dynamic works.

If Lindsay's biggest need is for belonging, and she has a belief that tells her, "I will be rejected," then the surgeon's critical outburst sends highly threatening signals to her brain, such as: "You're not included on this team. You do not belong here."

In seconds, Lindsay's prefrontal cortex floods with stress hormones, and she loses access to her knowledge, experience, and talents. Her need for belonging and her belief that she will be rejected have pre-disposed her to be shut down by this behavior. Whereas for Danelka and others, it is nothing more than an annoyance.

Intelligent Energy

We are conducting a research project that asks, "Are you smarter around some people than you are around others?" How would you respond if I asked you that question? When I ask it of groups of managers and leaders, the affirmative response is instantaneous and broad-scale. People nod their heads and say, "Oh you bet I am."

The survey then asks, "So if you're not as smart around some people – why is that so? What do they do to 'dumb you down'?" The most common responses are:

- They shoot me down before I can develop my thoughts
- They make me feel judged and evaluated
- They dismiss or discount my point of view
- They interrupt or talk over me
- They check out and disengage while I'm talking

Finally, respondents are asked, "If the people who make you feel smart get 100 percent of you, what percentage do the people get who dumb you down?" The average response is 45 percent. That means 55 percent of their brainpower is being left on the table. Does that ring true in your experience?

Brain science reveals why this is so. Cortisol (a stress hormone) constricts the brain and it resorts to rescue operations, proceeding along progressively rigid lines of thinking. It becomes reactive, extreme – binary. The brain abandons complex, future-based, generative thought as it scrambles for knee-jerk, black-and-white, simplistic solutions: "It's either now or never." "It's all or nothing." "You're either with me or against me." "I either have to fight you or flee from you."

When it comes to innovative energy, this is a problem, but it's also a big opportunity. As Woody and Szechtman have shown, if you can create the conditions in which people feel smart, you can help them ease out of this protective stance. And if you learn to help people recharge, they will have the energy that's needed to stand in the middle of business tensions, harmonize the competing priorities, and come up with breakthrough solutions that create value for their customers and co-workers.

Never forget, the world your employees live in moves at a frenetic pace. The mortgage, the kids, the commute, constant change, the unending flow of emails, and the annoying co-worker are all ingredients.

And (bad) stress not only impacts people's ability to think; it also kills their brain cells and shrinks their working memory. As you saw in chapter 4, managers can create a brain-friendly oasis in their departments with a way of being and simple energy conversations.

▶ CASE STORY
Unlocking Passion…and Extraordinary Sales

Robin is a good example of a manager who created these conditions in the midst of a tension-filled situation. She is a leader in charge of customer experience at a large financial institution. Her organization was wrestling with an issue that had a big impact on the customer experience: large tracts of customers weren't getting contacted.

Relationship managers (RMs) didn't have the capacity to make these calls, so head office had spearheaded an initiative to have customer service reps (CSRs) begin to do this. The initiative wasn't going well. The CSRs weren't making the calls.

The "story" at head office was this: "They either need more training or they're resistant to change." Robin had a different assumption – a more charitable one. She suspected that this was not about skills or resistance to change – that something else was getting in the way.

Robin and a colleague traversed the country, engaging in conversations to pull out the reality of the CSR. In the midst of this process, they asked us if there were different questions that would help them get to the real story of what was getting in the way for CSRs. We suggested they begin to ask the What Matters Most question. What happened next was astounding to all of us.

In one office was a CSR named Stacey who had been promoted to RM, but went back to her CSR role after having a baby so she could cut back on her travel. With her experience as an RM, Stacey was uniquely equipped to make the customer calls head office was expecting. Yet she wasn't doing it.

"Hmmm…" thought Robin. "Rule out 'they need more training' as the root cause of this situation."

When she began to ask Stacey what mattered most at work, she discovered two competing forces: Stacey's driving need to belong, and her belief that she would be rejected by her team if she became the organization's poster girl of customer calling.

Drawing out this need and this belief was the game-changer. Robin asked some coaching questions, enabling Stacey to reframe her situation and to begin to understand where her passion lay: serving her customers in a way that would not alienate her teammates.

Several weeks later, we got an email that I treasure to this day. Stacey's efforts to date had unlocked $7 million worth of business for the organization! Her ability to offer innovative energy had nothing to do with a lack of effort and everything to do with removing *interference*.

Energy Management Question

What's the biggest form of interference short-circuiting performance in our organization?

▶ ▶ ▶ Think Sticks, Not Carrots

9

Meet Needs, Not Scores

What's the brain science? Our brain makes decisions for emotional reasons, then justifies them with rational ones.

In chapter 3 we learned that our emotional brain tells us what is true. For example, if you don't feel valued by your boss, all the rational declarations and assurances in the world can never convince you it's real.

But the emotional brain goes far beyond defining reality: it acts as the inner arbiter in every decision your employees make. You may think reason is in the driver's seat when it comes to decision-making – that emotion sits quietly in the back and offers a wee preference here and a tiny opinion there. Nothing could be further from the truth.

As Gupta et al. write, twenty years of science shows that, when it comes to decision-making, emotion is in the driver's seat in ways we can hardly imagine. It's not that reason isn't involved. It clearly is. But logic and reason simply defend the conclusion that emotion has already decided.

Where does this show up at work? Emotion does have a logic to it, and this chapter points to that logic by revealing five emotional needs that drive employee decisions. These five needs feel as vital and urgent to employees as their need for oxygen. Once you understand this, one simple fact coalesces: employees' behaviors are their attempts to get their emotional needs met.

But current engagement strategies focus more on fixing scores than discovering the unmet needs of employees. And when those needs go unmet, they trigger a predictable reaction: people act out in unskillful ways to try to get their oxygen-like needs fulfilled.

Why does this matter? When we help managers to stop focusing on scores and help employees to skillfully get their needs met, it releases energy and pre-empts the interference that:

- Depletes energy
- Spawns frustration
- Short-circuits performance
- Erodes the employee experience
- Corrodes the customer experience
- Monopolizes the manager's time

Meeting employee needs generates a cycle of healthy decisions, reduced interference, and sustainable energy that makes engagement work for the long haul.

▶ ▶ ▶

The Lift

What if the emotional part of your brain were removed – that mercurial, frivolous, unpredictable part – and you could make choices based on cold, calculated logic alone. Would you be a better decision-maker? If, like many, you thought the answer was yes, you might want to think again.

For decades Antonio Damasio has worked with patients who have sustained impairment to the amygdala and/or the ventromedial prefrontal cortex. As Damasio has shown, when these two emotional centers of the brain fail to function, it produces a surprising outcome: patients are incapable of making the simplest of decisions.

In *Descartes' Error* he describes his patient Eliot, a young man who had surgery to remove a brain tumor but in the process ended up sustaining damage to his ventromedial prefrontal cortex (VPC).

Eliot had been a good husband and father, and an astute professional in a business firm. After his operation he became impulsive and lacking in self-discipline. He could not follow a schedule, perseverated on unimportant tasks while failing to recognize priorities, developed a passion for collecting "junk," and finally lost all his savings in a series of remarkably poor business judgments. His wife and family left, and he could no longer hold a steady job.

It is interesting to understand what changed about Eliot after his VPC was damaged. But it's even more interesting to understand what *didn't* change.

Eliot's I.Q. score still tested in the superior range. His perceptual ability, past memory, short-term memory, new learning, language, and the ability to do arithmetic were intact. More subtle tests of his ability to make inferences, to make estimates on the basis of incomplete knowledge, and even a personality test designed to pinpoint dysfunction, were passed with flying colors. Yet in the personal and social realm, Eliot was lost and baffled.

Our prefrontal cortex is a powerhouse when it comes to synthesizing, comparing, and processing bits of information. But it relies heavily on emotion to nudge it in the right direction. As Damasio puts it, "Without an emotional predisposition, one is left endlessly thinking of alternatives – giving everything equal weight whether they are relevant or not."

In an INET keynote speech, he explains this decision paralysis in terms of emotional lift. "The reason they can't choose is that they haven't got this lift that comes from emotion. It is emotion that allows you to mark things as good/bad or indifferent."

It's no surprise that emotion supplies this needed lift. "That's just what emotions do: affect motivations," says Medina, in *Brain Rules*.

And what does emotion base its decisions on? As a National Institutes of Health article puts it, "Decades of research have shown that the amygdala is involved in associating a stimulus with its emotional value. This tradition has been extended in newer work, which has shown that the amygdala is especially important for decision-making, by triggering autonomic responses to emotional stimuli."

R. L. Daft points out that our emotional system bases its decisions on "millions upon millions of experiences recorded in the nervous system from the moment of birth. We learn to sense, anticipate, and move toward pleasure and away from pain."

What Malcolm Gladwell called "thin-slicing," Damasio calls intuition. In *Descartes' Error* he points out that those logical steps that intuition seems to skip aren't necessarily skipped but "emotion delivers the conclusion so directly and rapidly that not much knowledge need come to mind."

This order of emotion first and logic second is supported by other scientists who have been able to witness this sequence during brain scans. According to Daft, "the choice-making regions of the brain are activated before people are aware they have made a choice. The rationale for the decision is constructed after the fact."

We like to see reason as our wise and learned teacher, logically assembling the facts and leading us to the wisest decision. But it's not. On the basis of his research on moral intuition, on how we make instantaneous judgments about others and the things they do, Jonathan Haidt says reason "works more like a lawyer or press secretary, justifying our acts and judgments to others."

What does all of this have to do with employee engagement? Every minute of the day, your employees are making decisions about how they will treat their work, their customers, one another, and you. They're basing those decisions on emotional reasons.

If those emotional reasons are triggered by unmet needs, their decisions could be producing unsavory outcomes. Simply put, learning how to identify and meet employee needs is an important management skill – as important for your employees as the air they breathe.

Five Driving Needs

Imagine your friend Bob is in the hospital and is having problems with his lungs. He's wearing an oxygen mask with a hose hooked up to an oxygen dispenser. One of Bob's relatives, Aunt Betty, bustles in to the room and pulls a chair up close to his bed. She is so intent on leaning in to talk to Bob that she fails to notice she's crimping his oxygen hose with her elbow. What do you suppose Bob's reaction might be, and how quickly do you think he would react?

Whether Bob gasps, "You're…crimping…my…hose!" or moves Betty's elbow off his hose, it's easy to imagine his reaction will be swift and decisive. Why? Because for Bob, oxygen is not a *preference* – it's a *driving need*.

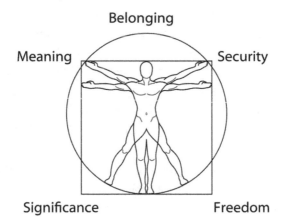

The oxygen metaphor is apt for the five driving needs that employees feel. Just as we constantly breathe – and act forcefully to keep breathing if it is endangered, so people are getting their driving needs met, right now, at this moment. Whether they are doing so in ways that are appropriate for others or themselves is another question entirely, but they are getting their needs met. Let's take a look at those five needs now.

We have a need to internally feel we *belong*. The research today is conclusive: the sense of belonging in strong social relationships is a greater predictor of one's physical health than whether they smoke, how much they exercise, or the quality of their diet. In short, isolation and loneliness will kill you faster than cigarettes. Why? Because belonging is like oxygen for human beings.

We also need to be able to internally feel we're *secure*. If a prankster wanted to shut down the entire financial sector of any major city, he could do so by simply placing one phone call: "I've placed a bomb on Bay Street."

People's productivity would instantly stop. All conversations would center on one thing. Our brains are highly reactive to any sense of danger, quickly switching off complex thought processes to keep us focused on survival. Security is not a mere preference or a desire – it is a driving need.

Freedom is another driving need. Think about freedom in the context of the uprisings we have seen in Egypt, Libya, and Syria. People's freedom can be suppressed only for so long before they will take unbelievable risks.

We also need a sense of *significance*. When someone does something that diminishes our self-worth or self-esteem, we react. Every one of us needs to be able to feel "I matter."

Finally, we need to feel a sense of *meaning*. Victor Frankl, the author of *Man's Search for Meaning*, discovered that he who has a strong *why* can bear any *how*. He observed that the people who were able to survive the atrocities of the prison camp he was in during World War II were the ones who had a piece of work they had to finish or a relationship they had to get back to. His conclusion? Man's search is not a search for happiness, but a search for meaning.

> *There's a difference between your employees' wants and their needs, and as a manager, it's important for you to be able to discern the difference. If you're curious about the difference, see the appendix "Wants vs. Needs."*

Families of Needs

Each of these five driving needs has a *family of needs* surrounding it – close cousins that bear the family resemblance but also bring their own distinct differences. As you look at these families of needs, you will probably begin to recognize the ones that are most important to you – or to your employees.

Belonging: acceptance and inclusion, identification and "insiderness," relatedness and intimacy, connection and fit.

Security: safety and protection, predictability and control, consistency and clarity, order and structure.

Freedom: autonomy and independence, mind and psychological "space," decision-latitude and support, variety and change.

Significance: respect and value, affirmation and acknowledgment, success and achievement, challenge and growth, efficiency and productivity, excellence and distinction, power and status.

Meaning: purpose and making a difference, understanding and connecting the dots, justice and fairness, altruism and service, creating positive change and inspiration, moral and ethical correctness.

Every score on your engagement survey is simply an indication of whether your employees feel these needs are being met. The challenge (and the opportunity) is that every employee has one or two needs that will matter most to them in any situation.

When you can draw out and acknowledge what matters most to an employee, interference can be removed and energy can be released in ways that make a positive difference in their lives.

When you don't invest the time to do this, every unmet need can show up in ways that suck the energy out of leaders, managers, and employees.

Unmet Needs Deplete Energy

As discussed in the previous chapter, our brain's sympathetic nervous system (SNS) is focused on fight or flight. When it's turned on, our body goes into energy consumption mode. In contrast, when our brain's parasympathetic nervous system (PNS) is turned on, we go into energy conservation mode.

Unmet needs trigger the fuel-guzzling SNS and quickly deplete people's energy. When you meet someone's needs, it turns off the SNS and activates the energy conservation mode of the PNS. This is energy management in a nutshell.

Which makes it clear that moving beyond engagement is the process of finding unmet needs and meeting them. Customers *need* value. Employees *need* structure and tools. Managers *need* strategic direction and effective systems. Leaders *need* results.

But when driving needs are not met, breakdowns occur. There's a name for unmet needs in marriage: "divorce." There's a name for unmet needs in manufacturing: "strike." There's a name for unmet needs in unstable countries: "revolution." There's a name for unmet needs between two countries: "war."

Situation	Fight Response	Flight Response
"I don't feel I belong"	People gossip about others or complain about changes in the system in an attempt to bond and cohere	People form exclusive cliques – isolating themselves from others
"I don't feel safe here"	People become controlling and micro-manage the process	People resort to CYA (cover your ass) behaviors and finger-pointing

Situation	Fight Response	Flight Response
"I'm feeling controlled"	People make decisions without others' input – counting on forgiveness rather than asking permission	People secretly bend rules and manipulate policies to get what they need
"I don't feel valued"	People take credit for others' ideas or bully their co-workers	People become political – quietly angling for position and power
"This is futile"	People become hypercritical – taking potshots at those who are trying to make a difference	People check-out in meetings, choosing not to offer their input

Why such forceful breakdowns? Because when you crimp someone's oxygen hose, he will act with terrific speed and unthinking aggression to restore the flow. People will do whatever it takes to get their driving needs met, whether through appropriate or inappropriate means.

Let's look at how this plays itself out in organizational life.

Our brains are wired to be highly reactive to any sense of threat. When people feel their needs are being threatened, they often act out with either a fight-or-flight response. But whether fight or flight, all of the behaviors in the list above have one thing in common. Can you spot what it is?

What do all these behaviors have in common? Perhaps you've guessed. Each one is an *unskillful expression of an unmet need*.

People who form cliques have a legitimate need for *belonging*. They are simply expressing this need unskillfully.

People who micro-manage don't do it because they're so buttoned down and organized. No, they're so buttoned down and organized because they have a valid human need for *security*. They are expressing this need in an unskillful way.

People who make decisions without others' input or who bend the rules to get what they need have a legitimate but poorly expressed need for *freedom*.

Similarly, people who bully or jockey for position have a legitimate need for *significance* and an inappropriate means of finding it.

People who take potshots at the company or disengage in meetings have a legitimate need for *meaning*. They hear about the vision and values but don't see anyone walking the talk. So they act out or pull within. Either way, they become very cynical and jaded.

Micro-managing. Insubordination. Bullying. Think of the organizational energy dissipated by these unskillful expressions. Grasping this one idea – *bad behavior is an unskillful expression of an unmet need* – is a useful way to neutralize these toxic elements and create more simplicity in organizational life.

Imagine if your first response to being micro-managed was to ask yourself, "What's the unmet need?" Asking yourself these four simple words triggers a vital behavior: a charitable interpretation.

If you ask yourself "What's the unmet need?" and recognize that the micro-manager may just have a need for more security, you can see if she needs more information up-front or more frequent check-ins – thereby lessening her need to micro-manage you.

Learning to think this way is important, because managers' and employees' driving needs often compete with those of others and become the source of unresolved conflict in an organization. Those who have a strong need for security can be threatened by those with a strong need for freedom – or other people's need for freedom can block still others' need for significance.

Getting Needs Met Skillfully

An effective engagement process should help people get their needs met skillfully. For example, let's say your employee has a driving need for significance. Recognition, praise, affirmation, and appreciation are what matter most to him. He might feel very vulnerable (and you uncomfortable) if he were to come to you and say, "I need more recognition and appreciation from you."

But he could come and say, "For me to do my best work, I need to know when I'm doing a good job. I also need to know when I'm not. Could we have that type of relationship?"

Most managers would welcome such an approach. It's a way to meet a need for recognition skillfully.

Let's say your driving need as a manager is for freedom and you have a new boss who is showing signs of being a micro-manager. It might not be the best possible approach for you to go to your boss and say, "I feel like you are micro-managing me." If you choose to take that route, don't tell anyone you got the idea from this book.

But you could go and say, "I do my best work when I have a fair amount of decision-making latitude. But if I were in your position, I'd want to know that nothing is going off the rails. I'm wondering if we could find the sweet spot – where I've got enough autonomy and you've got peace of mind knowing everything's under control?"

Getting your needs met skillfully expands your influence as a manager. Helping your employees get their needs met skillfully pre-empts the unskillful behaviors that produce interference, suck energy, ooze toxicity, and eat up all your time. Getting needs met early is one of the key ways of lifting the manager's burden.

But to do this well you have to understand something about the hidden marketplace and the currencies within it.

Trading with the Right Currency

Imagine an eminently wealthy and somewhat eccentric Hong Kong businessman who wants to have a vacation in Canada. What's eccentric about him is that he refuses to use credit cards and is obsessed with his own currency.

He comes to Canada with his family with a suitcase full of Hong Kong dollars. Predict how his vacation is going to go.

He and his family have finished a delicious (and expensive) meal in a fine local restaurant. The server brings the check and the businessman digs out his Hong Kong cash and lays it on top of the bill. How will the restaurateur respond? He might tell the gentleman to go and exchange his funny-looking money for Canadian cash. Or he might exchange it himself, offering a fraction of its true value.

My point is that *a currency that is highly prized by one person is meaningless to another*. What does this have to do with engagement?

Well, the relationships inside your organization function like a hidden marketplace.

The senior VP who cuts people off in meetings is trading in the currency of significance.

The newly promoted supervisor who still wants to be "one of the gang" and hesitates to hold her former co-workers accountable is trading in the currency of belonging.

The employee who asks question after irritating question in a meeting – drilling down to the fine granules until everyone's eyes roll – is trading in the currency of security.

International exchanges would grind to a halt but for the fact that they have established *equivalence* by which the value one party is offering feels equal to the value their trading partner is offering. Without equivalence, transactions would grind to a halt.

The same is true in the marketplace of work.

Access Points to the Five Driving Needs

Manager-employee relationship aside, saying, "I don't feel I belong" is challenging for anyone. Belonging may indeed be the driving need, but it's difficult to ask for it explicitly. So it's important for us to have comfortable access for employees, back doors, if you will, that feel non-invasive, are easy to approach, and lead us to the deeper driving needs.

For example, the employee who doesn't feel she belongs, due to bullying issues or cliquey teammate behaviors, can find it easier to talk about how she "fits" on the team or in her role.

Another employee might find it off-putting to say, "I'm not

feeling secure," but he could talk about clarity: agreeing on expectations or defining priorities.

To help you pull out employees' driving needs in a non-invasive but straightforward way, we've identified some access points: the shortest paths to the Five Driving Needs. Below, we express each of them as a positive statement that employees might say to themselves internally if they felt energized:

"I FIT"
"I'M CLEAR"
"I'M SUPPORTED"
"I'M VALUED"
"I'M INSPIRED"

The goal of these access points is twofold:

1 First, they are clues that can lead you to people's driving needs. For example, if an employee is saying he's unclear, it can be a sign that his need for security is going unmet.

2 Second, if you address one of these elements, there's a good chance you'll meet the underlying need as well. For example, create the conditions where an employee fits and there's a good chance you'll see that her need for belonging will be met as well.

Other examples of how these access points and the Five Driving Needs connect:

- In today's uncertain economic climate, when employees are clear about what's expected of them and how they're stacking up, there's a good chance they will feel more secure
- When an employee has all the support she needs – the tools, systems, backing, and authority to execute on her responsibilities – it tends to give her a sense of freedom
- When someone feels valued for her contribution, it fuels her feeling of significance
- When someone feels inspired about his work, the integrity

of his colleagues, and the purpose of his organization, it fosters a sense of meaning

Help Your Employee Get into Flow

Mihaly Csikszentmihalyi knows about what creates flow. He should: he's been researching it for forty years. Flow is a state of utter absorption in which the performer is so engaged that he or she loses track of time. Others refer to this state as being "in the zone," "in the pocket," or "in your sweet spot." One thing is clear: the flow state is an utterly energizing and attractively productive place to be.

The conditions that put people into a state of flow map closely to the elements of the Energy Check. I've translated Csikszentmihalyi's language into the terms we use when we train leaders.

What creates flow?

1 A healthy challenge: just the right amount of stretch ("I Fit")
2 Crystal-clear goals ("I'm Clear")
3 Minimal distractions and interference ("I'm Supported")
4 A rich flow of feedback ("I'm Valued")
5 Meaningful work ("I'm Inspired")

In the next chapter you'll see how managers can use a simple Energy Check that uses the five elements above to systematically identify opportunities to release more energy for employees.

Managers find it reassuring to know that the Energy Check is grounded in and founded on solid research such as Flow, Positive Psychology, Appreciative Inquiry, Strengths-Based Leadership, the Progress Principle, Emotional Intelligence, and Viral Change.

How Unmet Needs Can Serve Us

I belong to a community in which fellow athletes set significant goals for themselves and then train aggressively to achieve those goals. For example, my friend Geoff's ambition was to run the

Badwater race: 217 kilometers and 13,000 feet of steep, wicked hill-climbing through North America's hottest, most grueling desert.

How much motivation and discipline do you think a large, looming unmet need like this produces in an athlete? It is not small. This unmet need releases energy for the person to keep a strict diet, limit his alcohol intake, modify his sleep habits, and maintain a punishing workout schedule. But what happens the minute he crosses the finish-line? The energy and motivation evaporate. Why? The need has been met.

My friends have told me, "After I ran the Boston Marathon I didn't know what to do with myself. I had no motivation whatsoever. My diet and my training schedule went to pot. I need to set another goal for myself before I fall apart."

This shouldn't surprise us. According to humanist psychologist Abraham Maslow, *"only unmet needs motivate."* As soon as a need is met, we are no longer energized by it. Does this mean we should never meet people's needs? No, but it demands that we get very clear about the role of met needs and unmet needs – and how each can be used to help people perform.

In short, meeting needs removes interference and turns off the energy consumption mode of the sympathetic nervous system. Tapping in to unmet needs releases the high-performance hormones we noted in chapter 4. Leaders learn how to tap in to unmet needs by developing an amazing eye for potential.

The image below of the little man on the pogo stick can teach us something about potential and generating energy inside employees. As his weight comes down, the spring of the pogo-stick compresses and a millisecond later he finds himself sailing through the air.

What propels him? Energy. But where did it come from? When compressed, the spring is distorted from the shape it "wants" to be in: the atoms are scrunched into a distorted formation. It's very much like a pent-up, unmet need. That scrunching stores potential energy inside the spring. When the spring reaches its limit, all the atoms inside it rush to restore themselves to their undistorted state. That releases energy and our little man goes flying.

People in your organization are a bit like that spring. There is a state they "want" to be in. It's called their potential. Whenever they are not living up to their potential, they're in a distorted shape: every atom in their body is scrunched into an unnatural state. This creates tension – the tension between their current reality and their ideal state.

Good leaders develop an eye for potential. They see what someone can be and work with them to "release" them into that state. It's in the releasing process – moving from our distorted shape to our true state – that energy is generated.

The Antidote

There are three isms that threaten to destroy our planet: narcissism, individualism, and consumerism.

- "It's all about me"
- "I don't need anybody else"
- "I deserve to have whatever I want"

Sacrifice (old word) is the antidote to all three. In fact, there is no sustainable solution for our world apart from sacrifice.

You may be thinking, "Sacrifice. Heady stuff. Probably more in the domain of Mandela, Gandhi, Mother Theresa, and Jesus. It doesn't really apply to me."

But is that true? Think of an achievement that matters to you. My hunch is that it involved some sacrifice, choosing to set aside the comfort/pleasure of the moment for something more substantial.

Sacrifice is the code that's written right inside our lives on this planet. It's the reason you're here today. You may have been conceived in a moment of ecstasy, but you were sustained by a series of sacrifices.

I think the lack of sacrifice we complain about in our millennials started with baby boomers. We were unwilling to sacrifice connection/friendship with our kids, so we hesitated to teach sacrifice. "If I demand accountability, my kids won't want to have anything to do with me!" As a result, millennials face a big credibility chasm. How big?

A powerful study of millennial job seekers and veteran HR professionals done by Beyond.com reveals that 86 percent of millennials identified themselves as hard workers. Only 11 percent of HR professionals thought millennials would work hard.

This is a big PR problem. The tragedy is, you and I know many millennials who are hard-working, sacrificial, and loyal – but they've been tarred with a giant black brush because when managers look around there's sufficient evidence to suggest that the stereotype holds true. The Five Driving needs are an antidote.

The most important work you will ever do as a manager is not fixing the engagement scores in your department. It is helping a young employee (or any employee) expand the orbit of her contribution. It is:

- Accepting by creating belonging for others
- Protecting people who lack security

- Serving by helping those entrapped to experience freedom
- Caring by building people's sense of significance
- Challenging by inspiring people to find meaning

This is the essence of partnering. This is the check and balance: ensuring that the Five Driving Needs aren't about feeding the narcissism epidemic. This is what completes the cycle and ensures that the ecosystem of meeting needs is sustainable and doesn't begin to feed on itself.

This helps people get their highest needs met: the needs for contribution, meaning, and becoming a self-actualized human being. This is the antidote to narcissism, individualism, and consumerism. You simply cannot do the five things above without learning to sacrifice. As such, dealing skillfully with the five driving needs are the way we mature and evolve as human beings.

▶ CASE STORY
Making the Exchange

Consider Mandeep, who manages a young employee named Luke, who has a razor-sharp mind, loves to pump out results, and hates it when people waste his time. When Luke is in meetings, he uses dark humor and sarcastic zingers to shut people down if he feels they're not adding value. He thinks it's funny, but his teammates wilt under his caustic wit and refuse to offer any ideas in his presence.

Mandeep's currency is meaning. She's all about the highest good, contribution, and leaving a legacy. Later, she engages Luke in a conversation about his behaviors in the meeting and says, "How would it make you feel if people made comments like that to you?"

Oops. Wrong currency.

Luke says, "I'd love it. In fact I'd expect them to use sarcastic comments if my ideas don't have any value and I'm wasting people's time."

Mandeep instantly sees that there's no equivalence between her currency of meaning and his currency of significance. She course-corrects. "Luke, I believe you want to be seen as someone who is promotable – as a leader. Is that true?"

"Definitely."

"And I see that, too. To be seen as promotable, to be seen as a good leader, you have to be able to get the best out of people – right?"

"I agree."

"You have a fantastic wit and a great sense of humor, and I believe your intent is to use them to make sure the meeting is productive – to challenge people to bring their best ideas so we're not wasting time. Am I right?"

"You've got it."

"Well, the way it is right now, your sarcasm and dark humor are shutting people down. You're not getting the best stuff out of people so the meetings aren't as productive as they could be. My goal is to help this team contribute everything they can. So I want to work with you to see if you can use those talents in a way that not only lightens things up but also draws out people's best thinking."

At this point, Mandeep has found equivalence: they both want achievement, which for Luke is productivity and for her is contribution.

When you're partnering with your employees, the goal is to draw out what matters most to them (their currency), help them understand what matters most to you (your currency), and make the exchange, harmonizing your competing needs. Finding equivalent currencies is at the heart of what makes the organizational marketplace work.

Energy Management Question

What kinds of *unskillful expressions* are most prominent here and what are the unmet needs prompting them?

▶ ▶ ▶ Meet Needs, Not Scores

10

Challenge Beliefs, Not Emotions

What's the brain science? Our brain does not allot us the resources to do something until we believe we can do it.

Where does this show up at work? Engagement initiatives don't stall because people don't care or because people aren't good. They stall because people are low on self-efficacy: they lack the agency to move things forward.

Fortunately there's a cure for low self-efficacy. It lies hidden in an unexpected place: our beliefs. It's not our capability but *our belief in our capability* that produces self-efficacy. All sorts of capable people fail to do what they're capable of doing because of self-doubt, second-guesses, and frayed confidence; they simply don't believe they can.

Why does this matter? You want energized employees and you want that energy to be sustainable. But your employees and managers face obstacles, setbacks, moving targets, and roadblocks on a daily basis. When they lack agency, these challenges seem insurmountable.

Low agency dries up courage, stymies ingenuity, paralyzes risk-taking, short-circuits execution, and shuts down innovation.

Entropy exploits all this incapability and drains energy to a trickle. This infuses a deep, toxic cynicism into the organization.

▶ ▶ ▶

A Tale of Two Beliefs

When it comes to getting our needs met, each of us has two types of beliefs working within us: helpful beliefs that strengthen self-efficacy and unhelpful beliefs that cripple self-efficacy.

For example, if you believe, "I always get left holding the bag – I always get trapped in these situations," then your sense of agency in dealing with a micro-managing leader will be very low. Your chances of getting your need for freedom met are not great.

Like a Dragon's Den judge, your brain demands a sure bet and doles out resources only when it senses a high degree of confidence. So when you need to deal with your micro-managing boss and you do the pitch to your brain for some negotiation skills and it detects doubt and uncertainty, it sends you away empty-handed. As David DiSalvo has written, until you believe you can do something, your brain does not allot you the resources to do it.

You can probably see why this is so important. Your employees have beliefs about the oxygen-like needs we've talked about. They come to you pre-wired with beliefs concerning:

1 Whether or not they belong
2 How secure or in danger they are
3 Whether their freedom will be taken away or protected
4 How significant (or not) they are
5 How meaningful (or not) their work is

If they have a strong belief that they will be controlled, manipulated, or trapped by leaders (freedom beliefs) then the odds are very high that they might misinterpret your guidance as micro-managing.

Note: When I talk about beliefs, I'm not referring to religious beliefs. I'm exploring the larger body of beliefs each of us operates from that may include but is not confined to our religious beliefs. Not that it matters for our discussion, because the brain makes no such distinction. As Harris et al. write, "fMRI testing has revealed that participants' beliefs about the existence of Jesus Christ and the existence of the Eiffel tower light up the same area, 'the ventromedial prefrontal cortex,' an area important for self-representation, emotional associations, reward, and goal-driven behavior."

You may find it helpful to substitute a different word for beliefs – like assumptions, perceptions, rules, conclusions, filters, or lenses through which we view life. Pick your nomenclature and let's explore this brain-science truth, starting with an entity that is as deadly as it is ubiquitous: stress.

Stress: Demon or Demonized?

Stress has been demonized in today's workplace, but it's bad stress, not eustress (good stress), that short-circuits performance and burns people out. And a study by Keller et al. of 28,000 American adults over an eight-year period revealed that it's not stress but what you believe about stress that will kill you.

The study measured people's stress levels in the past year (Low-Medium-High) and then determined their beliefs about stress ("stress is harmful" vs. "stress is just how I gear up for a challenge").

Eight years later they used public health records to find out who died. The outcomes:

- People who experienced a lot of stress had a 43 percent greater chance of dying
- But this was true only for those who believed that stress is harmful – a threat to their health
- People who experienced a lot of stress but did not believe stress was harmful had the lowest risk of dying – including people who had relatively little stress

Researchers estimated that during the eight-year period of the study, 182,000 Americans died not from stress but from the belief that stress was bad for them. How could belief kill 182,000 Americans? Simply put, stress, without the resources to deal with it, is lethal. And your brain decides whether you get those resources or not.

Here's how DiSalvo describes this phenomenon in *Brain Changer*: "Despair is the belief that our situation cannot improve, and when we embrace this form of belief our brain responds by diverting energy away from action to improve our circumstances (because we believe it's hopeless) and into an eddy of negative rumination, fueling the downward spiral."

Hope, he says, is "the belief that our situation can and will improve no matter what, and when we fully embrace it, our brain responds with a deluge of mental energy to enable reaching the hopeful outcome."

DiSalvo adds, "You will not tap into the full array of mental resources available to you unless you hold yourself accountable for how much you genuinely believe you'll succeed."

In chapter 4 we explored the high-performance hormones the brain can make available to us and how these resources power up prioritization, concentration, goal-orientation, self-regulation, creativity, and intuition.

Now let's look at a study, by Jamieson et al., that connects the dots between "it's your belief about stress that kills you" and "the brain releases resources when it detects belief." Not surprisingly, the study explored how the participants' view of stress affected their perception of resources at their disposal.

Participants in the study were instructed to reappraise or rethink stress arousal as a functional response. Then they were submitted to acute stress. Reappraisal of the stress response produced a significant result in participants: a greater awareness of the resources available to them.

It's not difficult to see, then, how those in the stress study who believed that "stress is the natural way my body gears up for a challenge" would fare better (die in lower numbers) than those who believed that stress is harmful: their brains detected the belief that "stress is manageable" and unlocked the resources they needed to deal with it.

Researchers have shown a hidden treasure in stress. Von Dawans et al. write that it "triggers social approach behavior, which operates as a potent stress-buffering strategy in humans, thereby providing evidence for the tend-and-befriend hypothesis."

When you look at employee engagement through this lens, things begin to fall into place. Employees experience stress in terms of setbacks, roadblocks, time constraints, priority shifts, information overload, and change fatigue. If their belief is "stress is harmful," they can't even see or access the resources at their disposal, because the brain refuses to squander expensive processing power on a losing proposition.

But adopting a positive, *stress is helpful* belief unlocks the resources that improve performance. Here's how Crum et al. put it:

> People can be primed to adopt a stress-is-enhancing mindset, which can have positive consequences relating to improved health and work performance…People may not need to focus single-mindedly on reducing their stress. The message of this research is ultimately a positive one: eliciting the enhancing aspects of stress (as opposed to merely preventing the debilitating ones) may be, in part, a matter of changing one's mindset.

Work-life balance is venerated as the unquestioned right of every employee. Stress-reduction is the common promise. But it's bad stress, not eustress, that short-circuits performance and burns people out. Good stress energizes high performance, creates focus for remarkable results, and enables employees to thrive at work. Failing to understand this, leaders guarantee work-life balance and stress reduction as key engagement strategies.

Real cultural gains are made when managers:

- Help employees realize that it's what they believe about stress (not stress itself) that makes it harmful or helpful

- Set a clear expectation that seasons of (good) stress and recovery will be a normal part of the employee experience

Setting work-life balance and stress reduction up as twin aspirations guarantees that your employee engagement success will have a short shelf life. It creates an expectation in the minds of employees that can never be realized by real people in today's real world. This sets leaders up for failure and employees for cynicism.

Conversation Recalibrates Beliefs

Everything I said about conversation in chapter 4 comes to bear here. *Shifting beliefs* is not only the role of conversation, it is the innate power of conversation. In sum, conversation is powerful because it challenges the unhelpful beliefs of employees and nudges them to higher levels of self-efficacy.

Conversation is how partners hold out for each other's highest good. You use it to step in to each other's world, draw out what matters most, identify common ground, and pull out a bigger reality – a belief that's more reliable than either of your beliefs in isolation.

And Energy Check conversations, in particular, provide a rich opportunity for challenging unhelpful beliefs, both those of the employee and of the manager.

Holes in the Bucket

We've explored the Energy Check as a short, systematic way of identifying an employee's driving needs. Drawing those needs up to the surface and harmonizing them with your own needs requires some conversational skill. But while it's necessary to harmonize competing needs, it's not enough. A sustainable engagement process requires you to learn how to harmonize *competing beliefs* as well. Why? Identifying needs without also increasing the self-efficacy to get them met will only lead to frustration.

Let's say you have a new employee named Jerome who is highly reactive to criticism or negative feedback. After doing an Energy Check with Jerome, you are made freshly aware of his driving need for significance. For weeks afterward you go out of your way to offer him affirmation and recognition. If we use the "how full is your bucket" metaphor, you are frequently filling his bucket with praise.

A month goes by and you are puzzled when another employee informs you of Jerome's comment, "I never get any recognition for my work." How can this be? There must be gaping holes in Jerome's bucket.

Jerome's belief that "I don't matter" causes him to conclude that, though you're saying the words, you're saying them only because you're expected to – it's your job. Your recognition does not feel genuine to him.

Everything you say and do for Jerome is filtered through the "I don't matter" belief and causes him to feel ignored, passed over, and unappreciated. If Jerome's beliefs about his significance go unchallenged, you will exhaust yourself trying to meet his needs.

But let's back up a bit. How do beliefs like this imprint people in the first place?

Beliefs Are Burned In

Think back to Bob, your oxygen-starved friend in the hospital (chapter 8). He had the experience of Aunt Betty crimping his oxygen hose and was instantly imprinted with an intense feeling of panic. Now picture Bob in his bed a few days later as Betty enters his room and begins to approach his bed. What do you think is going on inside his head as she pulls her chair up beside him?

Apprehension. He begins to recoil and steer his oxygen hose away from her. Is Bob's behavior being driven by oxygen deprivation? No. It's being driven by a phantom threat: his perception about Betty's carelessness. His belief – "I'm not safe here" – is interacting with his need for security and driving his jumpy behavior.

Thousands upon thousands of emotional experiences have

been burned into your employees' amygdala, the computer chip inside their brains. Those experiences cause people to make conclusions about the world, themselves, and others: beliefs about the way things work.

These beliefs – how we perceive the world – trigger our emotions, and as we've seen, our emotions drive our decisions.

Welcome to organizational life. You take a new employee out for lunch and ask a few casual questions about his weekend, and he begins to bristle and shut down. What you don't know is that he had a manager in a previous organization he experienced as invasive and smothering. This experience was burned in to his emotional memory, and a belief was born: "Manager conversations are not safe. If I let conversation drift toward anything personal, managers will smother me with uncomfortable emotions."

This one simple belief, left unchallenged, will shut this employee off from connecting with managers who wish to support him. This belief has to do with security. Each of the other driving needs also has beliefs associated with it.

Managers Can Detect Beliefs

Employees seldom tell their manager what they believe (few even know), so we help managers learn to detect beliefs by listening to narratives and watching for behaviors (unskillful expressions).

Below is a small sample of unhelpful beliefs and the narrative and behaviors that can help you detect them.

Belief	Narrative	Behaviors
"I don't belong"	"...won't want me there..." "...never listen..."	• Forms cliques • Gossips/complains
"I am at risk"	"...can't trust that they'll..." "...could take advantage of..."	• Shifts blame/points finger • Decision-paralysis
"I'm trapped"	"...we have no choice now..." ..."never enough time..."	• Maverick/breaks rules • Withdraws/doesn't reply

Belief	Narrative	Behaviors
"I have no worth"	*"...why would they care..."* *"...can't seem to understand..."*	• Political posturing • Takes credit for others' ideas
"There's no purpose"	*"...it's not fair – it never is..."* *"...it won't make any difference..."*	• Cynical, jaded, checked out • Dark, sarcastic humor

Beliefs are powerful because they drive behaviors. Think of a conscious decision you made today. Perhaps you were sitting with a group of colleagues at lunch who were engaged in razor-sharp banter. You probably didn't consciously recognize it, but you were sitting somewhere along the gradient of one of the beliefs below and it was silently but surely affecting your decision to jump in or not to jump in and join the repartee.

"I belong"	*"I don't belong"*
"I'm safe"	*"I'm at risk"*
"I'm free"	*"I'm trapped"*
"I have worth"	*"I have no worth"*
"There's a purpose"	*"There's no purpose"*

If you concluded "I don't belong," you felt emotions of apprehension and self-doubt and you probably sat back and listened rather than trying to inject a comment. If your conclusion was "I belong," you felt emotions of connection and rapport and felt free to offer your wit.

Beliefs drive behaviors through the emotions. Your beliefs trigger your emotions, and your emotions tell you what you can and can't do. Because of this, you'll never nudge behaviors without recalibrating beliefs.

The story of Robin and Stacey in chapter 8 illustrates this. Robin not only identified Stacey's need for belonging, she went further and recognized her belief ("If I stand out I'll be rejected") and her emotion (fear of isolation). Robin was able to reframe the situation, helping Stacey see a way to excel at customer contact and safeguard the relationship with her co-workers. Stacey's emotions and behaviors changed when her beliefs were recalibrated. ("It's possible for me to do both.")

In *Immunity to Change* Robert Kegan shows that when an employee cognitively understands the change that needs to happen, emotionally agrees with it, and volitionally commits to it but fails to adopt the change, there is a hidden competing priority at work: a belief, a rule that is being protected at all costs. For Stacey that rule was "acceptance in the group comes before all else."

Acknowledge Emotions, Challenge Beliefs

We can't change the way we feel, but we can change what we believe. Our emotions are a physiological response, hardwired to our belief about a situation.

If an employee *believes* his boss is intentionally trying to make him look stupid in a meeting then emotions of anger and defensiveness will automatically register in his body and mind. He is not responsible for how he feels. He *is* responsible for how he responds and for what he believes, because his beliefs dictate his emotions.

That's why we help managers become skillful at acknowledging emotions and challenging beliefs when they do Energy Checks one-on-one or with their teams.

Challenging unhelpful beliefs is the only way I know of making energy sustainable. Without this skill, employees gravitate toward victimitis, powerlessness, and low self-efficacy.

This brain science truth brings us full circle. The first chapter showed that you can't have innovation without the power tools of the executive function of the brain. Now you know that it's not only energy but also belief that unlocks those metabolically expensive resources.

▶ CASE STORY
Detecting Beliefs to Sustain Energy

In a one-on-one Energy Check, Tanya indicates she is "red" (depleted) in the area of support.

"I never have enough time to do my job," she says. "I'm always scrambling."

Joy, her manager, is aware that Tanya, a relatively new employee, allows people to bring last-minute tasks to her without pushing back and managing expectations. But instead of jumping in with this rebuttal, Joy acknowledges Tanya's emotion.

"This sounds overwhelming. Tell me what it looks like in your day-to-day reality."

"I start my day with a list of things I want to do, and Michael shows up with a last-minute request for a PowerPoint deck. I start working on that, and Sheila comes and asks me to change the report I created for her the day before. She's always making little changes. By the end of the day I've hardly touched my to-do list. I answer emails till eleven at night, and the next day's the same thing all over again."

"You want to get stuff done, but you're stuck."

"You bet I am."

"What are your thoughts about how this could improve?"

"I think I need to go on a time management course."

"Did I see on your résumé that you've been to some time management training?"

"Yeah, but you get rusty. I need a refresher."

"Can I offer another perspective?"

"Sure."

"I've seen your to-do lists. You have your tasks all planned out and well prioritized. I don't actually believe time management is your issue. From what you've just said, I think managing others' expectations is what will make the difference for you."

"But I can't just say no to Michael or Sheila. I'm supposed to be supporting them."

"That's true. And how much of your best thinking can you give them when you're always playing catch-up?"

"There's that. But if I try to push back, I can see them coming to you and saying, 'She's not doing her job. She's not a team player.'"

"You're concerned about getting into trouble?"

"Uh-huh."

"That may be how it was for you in other places, but that's not the way we do things here. You not only have permission to push back when something's not working for you, it's an expectation that you will."

"Really."

"Really. And I will support you. I know you do good work for Michael, Sheila, and the rest of the team, and I don't want you feeling like you've never got enough time to do your job. Do you want me to talk to Mike and Sheila or do you want to give it a try?"

"Oh, I can do it, now that I know I won't get in trouble."

"You have my full backing."

"Thank you."

Even though the presenting information was, "I'm not capable – I need more skills," Joy detected that Tanya's belief was, "I'm not safe." That belief short-circuited the self-efficacy Tanya needed to push back and manage expectations.

Joy could have easily stepped in and had those push-back conversations herself with Michael and Sheila, but doing so would have left Tanya's agency stunted and undeveloped. Acknowledging Tanya's emotions sent a clear signal to her amygdala: "I am your confederate, not your enemy." This signal played a vital role in creating capacity in Tanya's mind, especially when her belief was broadcasting, "I'm not safe."

Once there's capacity, you can challenge the belief, which Joy did in a straightforward, supportive way.

Energy Management Question

How can we use conversation to build agency in our employees?

▶ ▶ ▶ Challenge Beliefs, Not Emotions

Conclusion

Moving beyond engagement is a matter of internalizing and walking out the ideas you've seen in this book:

- You need the discretionary effort and innovation of employees to produce sustainable results
- Managing energy, not engagement, is what powers up employees' effort and innovation
- Energy is released when you create a great employee experience
- But it's emotionally engaging experiences that unlock energy, not rational engagement, surveys, and programs
- Conversations are the way to remove interference and meet felt needs
- These conversations happen in a simple forum, a one-on-one or a team Energy Check
- Energy Checks harmonize not only competing needs but also competing beliefs, building higher levels of self-efficacy that unlock the brain's best resources
- Conversations draw out and pull out the treasure inherent in the inevitable tensions of the workplace
- Partnering is *how* you pull the treasure out of the tension

Managing energy rather than engagement is just smart business; it unlocks your employees' innovation, goodwill, and discretionary effort, allowing them to do the good they want to do and to take pride in producing results at work.

It really is up to you. You can build the conditions where employees flourish in the work that really matters. You can co-create a great employee experience that has people coming to the end of their careers saying to their friends and family, "Working with that manager – those were the best years of my life."

Which means you'll probably be saying they were your best years, too.

Appendix

Wants vs. Needs

Belonging

I want to be liked	*I need to feel included*
I want my ideas implemented	*I need to know my input is valued*

Security

I want to be in control	*I need to feel safe*
I want every detail	*I need to be clear*

Freedom

I want to be independent	*I need a sense of autonomy*
I want to be above the rules	*I need to challenge what doesn't make sense*

Significance

I want to be noticed	*I need to feel valued*
I want to be right	*I need to feel heard*

Meaning

I want to fix everything	*I need to contribute*
I want things to be the way they're supposed to be	*I need to know there's a purpose*

Sources

Introduction

Corporate Leadership Council (2004). "Driving Performance and Retention Through Employee Engagement." Retrieved from <http://www.mckpeople. com.au/SiteMedia/w3svc161/Uploads/ Documents/760af459-93b3-43c7-b52a-2a74e984c1a0.pdf>.

Chapter 1 *Manage Energy, Not Engagement*

Bade, S. (2010). "Cognitive Executive Functions and Work: Advancing from Job Jeopardy to Success Following a Brain Aneurysm. *Work: Journal of Prevention, Assessment & Rehabilitation* 36:389–98.

Baumeister, R. F. (2001). "Ego Depletion, the Executive Function, and Self-Control: An Energy Model of the Self in Personality." In Roberts, B. W. & Hogan, R. (Eds.) *Decade of Behavior*. Washington, DC: American Psychological Association, pp. 299–316.

Catalano, R., Goldman-Mellor, S., Saxton, K., Margerison-Zilko, C., Subbaraman, M., LeWinn, K., & Anderson, E. (2011). "The Health Effects of Economic Decline." *Annual Review of Public Health* 32:431–50.

CNN (April 22, 2005). "E-mails Hurt IQ More Than Pot." Retrieved from <http://www.cnn.com/2005/WORLD/europe/04/22/text.iq/>.

Drubach, D. (2000). *The Brain Explained*. Upper Saddle River, NJ: Prentice-Hall.

Gailliot, M. T. (2008). "Unlocking the Energy Dynamics of Executive Functioning: Linking Executive Functioning to Brain Glycogen." *Perspectives on Psychological Science* 3:245.

Holtzer, R., Shuman, M., Mahoney, J. R., Lipton, R., & Verghese, J. (2010). "Cognitive Fatigue Defined in the Context of Attention Networks." *Aging, Neuropsychology, and Cognition* 18:108–28.

PRNewsire.com (May 1, 2012). "Americans Crave Sleep More Than Sex, Says Better Sleep Council Survey." Retrieved from <http://www.prnewswire. com/news-releases/americans-crave-sleep-more-than-sex-says-better-sleep-council-survey-149697505.html>.

Van der Linden, D., Frese, M., & Meijman, T. F. (2003). "Mental Fatigue and the Control of Cognitive Processes: Effects on Perseveration and Planning." *Acta Psychologica* 113:45.

Van Ruysseveldt, J., Verboon, P., & Smulders, P. (2011). "Job Resources and Emotional Exhaustion: The Mediating Role of Learning Opportunities." *Work & Stress* 25:205–23.

Chapter 2 *Deliver Experiences, Not Promises*

Arbuthnott, G. W. & Wickens, J. (2007). "Space, Time and Dopamine." *Trends in Neurosciences* 30:62–69.

thebrain.mcgill.ca (n.d.). "How Drugs Affect Neurotransmitters." Retrieved from <http://thebrain.mcgill.ca/flash/i/i_03/i_03_m/i_03_m_par/i_03_m_par_nicotine.html>.

Heath, R. G. (1972). "Pleasure and Brain Activity in Man: Deep and Surface Electroencephalograms During Orgasm." *Journal of Nervous and Mental Disease* 154:3–18.

Heskett, J., Jones, T., Loveman, G., Sasser, E., & Schlesinger, L. (March/April 1994). "Putting the Service Profit Chain to Work." *Harvard Business Review* 72(2):164–74.

Knowledge.alliantz.com <http://knowledge.allianz.com/finance/behavioral_finance/?1819/save-more-tomorrow-a-guide-to-smarter-saving>.

Olds, J. (1956). "Pleasure Centers in the Brain." *Scientific American* 195(4):105–16.

_____ (1958). "Self-Stimulation of the Brain: Its Use to Study Local Effects of Hunger, Sex, and Drugs." *Science* 127:315–24.

Royal, M. & Agnew, T. (2012). "The Enemy of Engagement: Put an End to Workplace Frustrations and Get the Most from Your Employees." New York: Hay Group, pp. 219–20.

telenav.com (August 3, 2011). "Survey Finds One-third of Americans More Willing to Give up Sex Than Their Mobile Phones." <http://www.telenav.com/about/pr-summer-travel/ report-20110803.html>.

Volkow, N. D., Fowler, J. S., Wang, G. J., Baler, R., & Telang, F. (2009). "Imaging Dopamine's Role in Drug Abuse and Addiction." *Neuropharmacology* 56:3–8.

Wise, R. A. (2004). "Dopamine, Learning and Motivation." *Nature Reviews Neuroscience* 5:483–94.

Wood, P. B., Schweinhardt, P., Jaeger, E., Dagher, A., Hakyemez, H., Rabiner, E. A., Bushnell, M. C., & Chizh, B. A. (2007). "Fibromyalgia Patients Show an Abnormal Dopamine Response to Pain." *European Journal of Neuroscience* 25:3576–82.

Chapter 3 *Target Emotion, Not Logic*

Buckingham, M. & Coffman, C. (1999). *First, Break All the Rules: What The World's Greatest Managers Do Differently*. New York: Simon & Schuster.

Corporate Leadership Council (2004). "Driving Performance and Retention Through Employee Engagement." Washington, DC: Corporate Executive Board. Retrieved from <http://www.mckpeople.com.au/SiteMedia/w3svc161/Uploads/ Documents/760af459-93b3-43c7-b52a-2a74e984c1a0.pdf>.

De Becker, G. (1997). *The Gift of Fear: And Other Survival Signals That Protect Us from Violence*. New York: Dell Publishing.

Freeman, J., Stolier, R., Ingbretson, Z., & Hehman, E. (August 6, 2014). "Amygdala Responsivity to High-Level Social Information from Unseen Faces." *The Journal of Neuroscience* 34(32):10573–81.

Kouzes, J. M. & Posner, B. Z. (2003). *Encouraging the Heart: A Leader's Guide to Rewarding and Recognizing Others*. Hoboken, NJ: Jossey-Bass Publishers, pp. 9–10.

Leaf, C. (2008). *Who Switched Off My Brain?* Dallas, TX: Switch on Your Brain USA.

Rizzolatti, G. (March 11, 2011). <http://www.psychologicalscience.org/index.php/publications/observer/2011/march-11/reflections-on-mirror-neurons.html>.

Rock, D. (2009). *Your Brain at Work: Strategies for Overcoming Distraction, Regaining Focus, and Working Smarter All Day Long*. New York: HarperCollins Publishers.

Bergland, C. (2012). "The Neurochemicals of Happiness." *Psychology Today*, last modified November 29, 2012 <http://www.psychologytoday.com/blog/the-atheletes-way/201211/the-neurochemicals-happiness>.

Goleman, D. (1998). *Working with Emotional Intelligence*. New York: Bantam Books.

Goleman, D., Boyatzis, R., & Mckee, A. (2013). *Primal Leadership: Unleashing the Power of Emotional Intelligence*. Boston: Harvard Business Review Press, p. 7.

Hallowell, E. (January–February 1999). "The Human Moment at Work." *Harvard Business Review* 77(1):58–66.

Iacoboni, M., Molnar-Szakacs, I., Gallese, V., Buccino, G., Mazziotta, J. C., & Rizzolatti, G. (2005). "Grasping the Intentions of Others with One's Own Mirror Neuron System." *PLOS Biology* 3(3):529–35.

Keysers, C., Wicker, B., Gazzola, V., Anton, J. L., Fogasi, L., & Gallese, V. (2004). "A Touching Sight: SII/PV Activation During the Observation and Experience of Touch." *Neuron* 42(2):335–46.

Kishida, K. T., Sandberg, S. G., Lohrenz, T., Comair, Y. G., Sáez, I., Phillips, P. E. M., & Montague, P. R. (August 4, 2011). "Sub-Second Dopamine Detection in Human Striatum." *Public Library of Science Journal*. DOI: 10.1371/journal.pone.0023291 <http://research.vtc.vt.edu/news/2011/oct/27/dopamine-release-human-brain-tracked-microsecond-t/>.

McGaugh, J. L. & Roozendaal, B. (2002). "Role of Adrenal Stress Hormones in Forming Lasting Memories in the Brain." *Current Opinion in Neurobiology* 12:205–10. Published online February 21, 2002.

Panksepp, J. (1998). "Affective Neuroscience: The Foundations of Human and Animal Emotions" <http://mybrainnotes.com/serotonin-dopamine-epinephrine.html>.

Rizzolatti, G., Fadiga, L. Gallese, V., & Fogassi, L.(1996). "Premotor Cortex and the Recognition of Motor Actions." *Cognitive Brain Research* 3:131–41.

Rock, David (2007). *Quiet Leadership: Six Steps to Transforming Performance at Work*. New York: HarperCollins Publishers.

Wicker, B., Keysers, C., Pially, J., Royet, J-P, Gallese, V., & Rizzolatti, G. (October 30, 2003). "Both of Us Disgusted in My Insula: The Common Neural Bases of Seeing and Feeling Disgust." *Neuron* 40(3):655–64.

Zak, P. J. (2012). *The Moral Molecule*. New York: Dutton / Penguin Group (USA).

Chapter 5 *Seek Tension, Not Harmony*

Arnsten, A. (2009). "Stress Signalling Pathways That Impair Prefrontal Cortex Structure and Function." *Nature Reviews Neuroscience* 10(6):410–22.

Fritz, R. (1989). *The Path of Least Resistance: Learning to Become the Creative Force in Your Own Life*. New York: Ballantine Books.

Ghacibeh, G. A., Shenker, J. I., Shenal, B., Uthman, B. M., & Heilman, K. M. (2006). "Effect of Vagus Nerve Stimulation on Creativity and Cognitive Flexibility." *Epilepsy & Behavior* 8:720.

Girotra, K. & Netessine, S. (March 27, 2013). "Liberate Your Employees and Recharge Your Business Model." HBR Blog Network.

Kim, C. & Mauborgne, R. (2005). *Blue Ocean Strategy: How to Create Uncontested Market Space and Make Competition Irrelevant*. Boston: Harvard Business Review Press.

Lencioni, P. (2009). *Death by Meeting: A Leadership Fable...About Solving the Most Painful Problem in Business*. San Francisco: Jossey-Bass.

Levy, N. J. (1961). "Notes on the Creative Process and the Creative Person." *Psychiatric Quarterly* 35:67.

Rock, D. & Page, L. (2009). *Living with the Brain in Mind: Foundations for Practice*. Hoboken, NJ: John Wiley & Sons.

Seaward, B. L. (2011). *Managing Stress: Principles and Strategies for Health and Well-Being*. Burlington, MA: Jones & Bartlett Learning, p. 41.

Siegel, D. (2010). *The Mindful Therapist: A Clinician's Guide to Mindsight and Neural Integration*. New York: W. W. Norton & Company.

Chapter 6 *Practice Partnering, Not Parenting*

Martin, R. (2002). *The Responsibility Virus: How Control Freaks, Shrinking Violets and the Rest of Us Can Harness the Power of True Partnership*. New York: Basic Books, p. 50.

Rock, D (2009). "Managing with the Brain in Mind." *Strategy + Business* 56:7.

Chapter 7 *Pull Out the Backstory, Not the Action Plan*

Amit, D. J. (1989). *Modeling Brain Function: The World of Attractor Neural Networks*. New York: Cambridge University Press, p. 10.

Brothers, L. (1997). *Friday's Footprint: How Society Shapes the Human Mind.* New York: Oxford University Press.

_____ (1997 and 2001). *Mistaken Identity: The Mind-Brain Problem Reconsidered.* New York: State University Press of New York.

Franks, D. (2010). *Neurosociology: The Nexus Between Neuroscience and Sociology.* Springer Science and Business Media.

Gregory, S. (1999). "Navigating the Sound Stream of Human Social Interaction." In Franks, D. & Smith, T. (Eds.). *Mind, Brains and Society: Toward a Neurosociology of Emotion.* Stamford, CT: JAI Press, pp. 254–56.

Herrero, Leandro. (2011). *Homo Imitans: The Art of Social Infection – Viral Change in Action.* High Wycombe, UK: Chalfont Project T/A Meeting Minds Publishing.

Lieberman, M. (2013). *Social: Why Our Brains Are Wired to Connect.* New York: Crown Publishers, pp. 9, 19.

Nasuto, S-J, Bishop, J. M., & de Meyer, K. (2014). "Communicating Neurons: A Connectionist Spiking Neuron Implementation of Stochastic Diffusion Search." *Neurocomputing* 72:707.

National Institutes of Health, U.S. Department of Health and Human Services. (n.d.) "Neurons, Brain Chemistry, and Neurotransmission." Retrieved from <http://science.education.nih.gov/supplements/nih2/addiction/guide/lesson2-1.htm>.

Chapter 8 Think Sticks, Not Carrots

Amabile, T. & Kramer, S. (2011). *The Progress Principle: Using Small Wins to Ignite Joy, Engagement, and Creativity at Work.* Boston: Harvard Business Review Press, pp. 7, 92–93.

Basile, B., Bassi, A., Calcagnini, G., Strano, S., Caltagirone, C., Macaluso, E., Cortelli, P., & Bozzali, M. (2013). "Direct Stimulation of the Autonomic Nervous System Modulates Activity of the Brain at Rest and When Engaged in a Cognitive Task." *Human Brain Mapping* 34:1605.

Baumeister, R. F., et al. (2001). "Bad Is Stronger Than Good." *Review of General Psychology* 5(4):323.

De Becker, G. (1998). *The Gift of Fear: And Other Survival Signals That Protect Us from Violence.* New York: Dell Publishing.

Frost, P. J. (2003). *Toxic Emotions at Work and What You Can Do About Them.* Boston: Harvard Business Review Press, p. 35.

Mizuno, K., Tanaka, M., Yamaguti K., Kajimoto, O., Kuratsune, H., & Watanabe, H. (2011). "Mental Fatigue Caused by Prolonged Cognitive Load Associated with Sympathetic Hyperactivity." *Behavior and Brain Functions* 7: 6.

Pearson, C. & Porath, C. (2009). *The Cost of Bad Behavior: How Incivility Is Damaging Your Business and What You Can Do About It.* Hay Group, p. 31.

Royal, M. & Agnew, T. (2012). *The Enemy of Engagement: Put an End to Workplace Frustration – and Get the Most from Your Employees.* New York: Amacom.

Woody, E. Z. & Szechtman, H. (2011). "Adaptation to Potential Threat: The Evolution, Neurobiology, and Psychopathology of the Security Motivation System." *Neuroscience and Biobehavioral Reviews* 35:1024.

Chapter 9 *Meet Needs, Not Scores*

Beyond.com (May 28, 2013). <http://about.beyond.com/press/ releases/20130528-Beyondcom-Survey-Uncovers-How-Veteran-HR-Professionals-Really-Feel-about-Job-Seekers-from-Millennial-Generation>.

Csikszentmihalyi, Mihaly (2004). *Flow: The Psychology of Optimal Experience.* New York: Penguin.

Daft, R. L. (2007). *Organizational Theory and Design.* Mason, OH: Thomson, pp. 25, 55.

Damasio, A. (1994). *Descartes' Error: Emotion, Reason, and the Human Brain.* New York: G. P. Putnam's Sons.

_____ . (2003). *Looking for Spinoza: Joy, Soprrow, and the Human Brain.* New York: G. P. Putnam's Sons.

_____ . (November 26, 2012). "Somatic Marker Hypothesis 1." INET Keynote <http://www.youtube.com/watch?v=Q_KUv8e-6vy>.

Frankl, V. (2006). *Man's Search for Meaning: An Introduction to Logotherapy.* Boston: Beacon Press.

Gupta, Rupa, et al. (2001). "The Amygdala and Decision Making." *Neuropsychologia* 49(4).

Haidt, Jonathan (March 23, 2012). <http://www.nytimes. com/2012/03/25/books/review/the-righteous-mind-by-jonathan-haidt. html?pagewanted=all&_r=0>.

Maslow, A. (1943). "A Theory of Human Motivation." *Psychological Review* 50:370.

Medina, J. (2008). *Brain Rules: 12 Principles for Surviving and Thriving at Work, Home, and School*. Seattle: Pear Press, p. 215.

National Institutes of Health <http://www.ncbi.nlm.nih.gov/pmc/articles/PMC3032808/>.

Chapter 10 *Challenge Beliefs, Not Emotions*

Crum, A. J., Salovey, P., & Achor, S. (2013). "Rethinking Stress: The Role of Mindsets in Determining the Stress Response." *Journal of Personality and Social Psychology* 104:729.

DiSalvo, D. (2013). *Brain Changer: How Harnessing Your Brain's Power to Adapt Can Change Your Life*. Dallas, TX: Benbella Books, p. 83.

Harris, S., Kaplan, J. T., Curiel, A., Bookheimer, S. Y., Iacoboni, M., et al. (2009). "The Neural Correlates of Religious and Nonreligious Belief." *PLoS ONE* 4(10):e7272. doi:10.1371/journal.pone.0007272.

Jamieson, J. P., Nock, M. K., & Mendes, W. B. (2012). "Mind Over Matter: Reappraising Arousal Improves Cardiovascular and Cognitive Responses to Stress." *Journal of Experimental Psychology: General* 141(3):420.

Kegan, R. & Lakey, L. L. (2009). *Immunity to Change: How to Overcome It and Unlock the Potential in Yourself and Your Organization*. Boston: Harvard Business Review Press.

Keller, A., Litzelman, K., Wisk, L. E., Maddox, T., Cheng, E. R., Creswell, P. D., & Witt, W. P. (2012). "Does the Perception That Stress Affects Health Matter? The Association with Health and Mortality." *Health Psychology* 31:677–84.

Von Dawans, B., Fischbacher, U., Kirschbaum, C., Fehr, E., & Heinrichs, M. (July 1, 2012). "The Social Dimension of Stress Reactivity: Acute Stress Increases Prosocial Behavior in Humans." *Psychological Science* 23(7):829.

 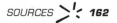

Index

Freeman, Jonathan et al., 36
Fritz, Robert, 68
frustration, 108
functional magnetic resonance imaging (fMRI), 49, 140

Galliot, Michael, 10
gap
    between customer experience and employee experience, 67
    between manager intention and employee feeling, 38, 44
Ghacibeh, Georges, 68
Gift of Fear, The (Gavin de Becker), 109
gift box story, 19–20
Girotra, Karan, 67
Gladwell, Malcolm, 122
Goleman, Daniel et al. (*Primal Leadership*), 47, 48
Great Place to Work Institute, 104, 106
Gregory, S.W., Jr., 101
Gupta et al., 119

Hadley and MacKay, 37
Haidt, Jonathan, 122
Hallowell, Edward, 53
Harris et al., 140
Hay Group, 108
HBR Blog Network, 67
Heath, Robert G., 23
Henry, Brian, 104
Herrero, Leandro (*Homo Imitans*), 101
high-performing employees, 25
highest good, 93–94, 95, 136, 143
Homo Imitans (Herrero), 101
hormones, high-performance, 5, 45, 46, 47, 50, 133, 141
    cortisol, 77–78, 116
    dopamine, 23–24, 51–53, 55, 56, 57
    and five driving needs, 57
    oxytocin, 51, 53–55, 57
    seratonin, 51, 53, 55–56, 57
HR professionals
    and millennials, 135
Hutchison, William, 49

Immunity to Change (Kegan), 147
insubordination, 128
interference to performance, 113–114
    costs of, 115
    "dumbing down," 116
    human energy depleted by, 114
    pre-empted by meeting employee needs, 120
    psychological needs and, 116
    removal of, 6, 118, 133
    as trigger to sympathetic nervous system, 114

Also by Brady G. Wilson

- Use *Beyond Engagement* and Brady's other books as a professional personal-development tool throughout your organization. Contact Juice Inc. to take advantage of substantial discounts when you purchase multiple copies.

- Brady is available for keynotes, training, and consulting. He is a highly animated, intensely pragmatic presenter, trainer, and thought leader. As a keynote presenter, Brady shares practical tools and the know-how to help business leaders step in to life's grittiest tensions: creating an audience experience that moves concepts from theory to application.

- Visit the Juice Resource Centre at www.juiceinc.com for information about webinars, speaking engagements, white papers, videos, articles, toolkits, and upcoming events.

Get refreshed. Get real results.
Get Brady Wilson to speak at your next event!

Connect with Brady:
linkedin.com/in/bradyjuiceinc | twitter.com/bradyjuiceinc

1-888-822-5479 | brady@juiceinc.com | www.juiceinc.com

BEYOND ENGAGEMENT

Energized Employees
Get Better Results

The Beyond Engagement training program helps organizations learn how to create an environment that is very highly engaged. How? By providing tools and resources to help leaders shift from managing employee engagement to releasing and optimizing sustainable employee energy.

What leaders are saying about the Beyond Engagement training program...

"The whole [Beyond Engagement] experience exceeded my expectations. It left me wanting more!"

"This was SIMPLE TO IMPLEMENT and it is saving me time!"

"My employees are solving their own challenges. They ARE OWNING THEIR ENGAGEMENT and asking for Energy Checks."

"This bolted right into the other training I have taken and made it work more effectively!"

"We are getting at performance issues in a non triggering way."

"[The] Beyond Engagement [program] gave us the tools and skills to help us describe how we feel about things – particularly things getting in the way of our energy levels and engagement...Before, we didn't know how to put this stuff into words. Beyond Engagement has given us that language."

"Partnering vs. parenting has made a huge positive impact with how I now lead my team."

"The Energy Checks, when done correctly, are imperative to keeping communication open and the team energized and motivated."

Engage with us at Juice Inc. to discover proven skills, tools, and training strategies for your organization.

Connect with Juice Inc.:

1-888-822-5479 | info@juiceinc.com | www.juiceinc.com

linkedin.com/company/juiceinc-

twitter.com/juiceinc

facebook.com/JuiceInc